Alternative Security

Living Without Nuclear Deterrence

Alternative Security

Living Without Nuclear Deterrence

EDITED BY

Burns H. Weston
THE UNIVERSITY OF IOWA

Westview Press
BOULDER, SAN FRANCISCO, & OXFORD

This book may be cited as

Burns H. Weston (ed. & contrib.), *Alternative Security:*
Living Without Nuclear Deterrence (Boulder, CO: Westview, 1990)

The epigraph on p. vii is from Bob Marley's "Redemption Song" (Bob Marley Music Ltd. – Almo Music Corp. ASCAP), produced in 1980 by Island Records Ltd. Used by permission.

Published in 1990 in the United States of America by Westview Press, Inc., 5500 Central Avenue, Boulder, Colorado 80301, and in the United Kingdom by Westview Press, Inc., 36 Lonsdale Road, Summertown, Oxford OX2 7EW

Library of Congress Cataloging-in-Publication Data
Alternative security : living without nuclear deterrence / edited by
    Burns H. Weston
        p.   cm.
    Includes bibliographical references and index.
    ISBN 0-8133-0629-9 (hardcover) – ISBN 0-8133-0630-2 (pbk.)
    1. Security, International. 2. Peace. 3. Nuclear warfare.
I. Weston, Burns H., 1933-   .
JX1952.A6632 1990
327.1'7 – dc20   90-12680
                                                                              CIP

Printed and bound in the United States of America

 The paper used in this publication meets the requirements of the American National Standard for Permanence of Paper for Printed Library Materials Z39.48-1984.

10 9 8 7 6 5 4 3 2 1

To the memory of

Ted Kjaer

and to all his family

with love

Emancipate yourself from mental slavery,
None but ourselves can free our minds.

—Bob Marley
"Redemption Song"

Contents

Acknowledgments

All scholarly endeavors borrow from others. In this case the debt is especially large. Without the hard work and good will, under very trying circumstances at times, of the authors who contributed generously to this volume, manifestly the book could not have been realized. To them I extend my primary thanks.

I am grateful as well to my students and colleagues for helpful dialogue and feedback. I especially want to thank my research assistants, past and present: Michael R. Hall, James W. McCormick, and James C. Wherry. They gave of themselves generously and creatively.

Thanks also to my daughter, Rebecca B. Weston, for the Bob Marley quote. Perfect!

Mary E. Sleichter, my secretary, displayed wonderful patience and, as usual, brought much-needed assistance to the repeated preparation of the manuscript. I thank her for her kindness, as I do also my friends Gregory P. Johnson for his expert computer assistance and Joan M. Kjaer who, in the early stages, graciously attended to my progress with sensitive eye and ear.

I wish to thank, too, Dean N. William Hines of The University of Iowa College of Law for his continuing support of my work, sometimes outside the mainstream of the law school world, as in this instance. I am especially indebted to him for the funding that helped to underwrite research assistance and miscellaneous production costs. These generosities are never taken for granted, never unappreciated.

Finally, I want to thank Miriam Gilbert, Jennifer Knerr, and Martha Leggett, my editors. It has been a pleasure to work with them.

Burns H. Weston
Iowa City, Iowa

About the
Editor and Contributors

WILLIAM SLOANE COFFIN is President of Sane/Freeze: Campaign for Global Security in Washington, D.C. From 1977 to 1987, he was Senior Minister of the Riverside Church in New York City and, before that, for eighteen years, Chaplain of Yale University. Reverend Coffin is also the author of *Once to Every Man* (New York: Atheneum, 1977), *The Courage to Love* (San Francisco: Harper and Row, 1982), and *Speaking the Truth in a World of Illusions* (San Francisco: Harper and Row, 1985).

WARREN F. DAVIS is President of Davis Associates, Inc., a firm specializing in the development and marketing of advanced scientific software, based in Newton, Massachusetts. For about ten years, he was employed by the United States defense industry, having been drafted into such work, with secret security clearance at age sixteen while still in high school, in response to the 1957 launching of the Soviet Sputnik. A holder of both undergraduate and Ph.D. degrees in theoretical physics from the Massachusetts Institute of Technology, Dr. Davis is co-founder of High Technology Professionals for Peace and has written and lectured widely on professional issues in defense employment.

LLOYD J. DUMAS is Professor of Economics and Political Economy at The University of Texas (Dallas). He is the author of *The Overburdened Economy* (Berkeley, CA: University of California Press, 1986) and editor and coauthor of *Making Peace Possible* (Elmsford, NY: Pergamon, 1989) and *The Political Economy of Arms Reduction: Reversing Economic Decay* (Boulder, CO: Westview, 1982). Professor Dumas also has lectured widely and authored more than sixty articles published in nine languages in books and journals of six disciplines as well as in the popular press.

ROBERT C. JOHANSEN is Director of Graduate Studies at the Institute for International Peace Studies at the University of Notre Dame. A past president of the Institute for World Order in New York City, Dr. Johansen is also a member of the Board of Directors of the Arms Control Association (Washington, D.C.), a contributing editor of the *World Policy Journal*, and the author of *Toward a Dependable Peace: A Proposal for an Appropriate Security System* (New York: Institute for World Order, 1978), *The National Interest and the Human Interest: An Analysis of United States Foreign Policy* (Princeton, NJ: Princeton University Press, 1980), and *Toward An Alternative Security System: Moving Beyond the Balance of Power in the Search for World Security* (New York: World Policy Institute, 1983). Dr. Johansen also edited *The Nuclear Arms Debate: Ethical and Political Implications* (Princeton, NJ: Center for International Studies, Princeton University, 1984).

THOMAS F. LYNCH, III, is Associate Professor of International Relations and Security Studies in the Department of Social Sciences at the United States Military Academy in West Point, New York. He currently is a candidate for a doctoral degree in International Relations/National Security Studies at the Woodrow Wilson School of Public and International Relations at Princeton University. He also is an active duty Armor Captain in the United States Army, who successfully completed, during 1983-88, company-level commands of a Tank Company in the Republic of Korea and a Border Cavalry Troop in the Federal Republic of Germany.

BRUCE RUSSETT is Dean Acheson Professor of International Relations and Political Science at Yale University and has held visiting appointments in Belgium, England, Israel, and the Netherlands. Professor Russett also is Editor of the *Journal of Conflict Resolution* and a past president of the International Studies Association and the Peace Science Society (International). The latest of his nineteen books is *Controlling the Sword: The Democratic Governance of National Security* (Cambridge, MA: Harvard University Press, 1990).

BURNS H. WESTON is Bessie Dutton Murray Professor of Law at The University of Iowa, specializing in international law and affairs.

A Senior Fellow of the World Policy Institute (New York City), a Fellow of the World Academy of Art and Science, and a member of the American Bar Association's Standing Committee on World Order Under Law and the Board of Editors of the *American Journal of International Law*, he also is active in the Lawyers' Committee on Nuclear Policy and the International Association of Lawyers Against Nuclear Arms. Professor Weston's many publications include *Toward Nuclear Disarmament and Global Security: A Search for Alternatives* (Boulder, CO: Westview, 1984), *Human Rights in the World Community: Issues and Action* (Philadelphia: University of Pennsylvania Press, 1989), and *International Law and World Order: A Problem-Oriented Coursebook* (St. Paul, MN: West Publishing, Second Edition, 1990).

RALPH K. WHITE, Emeritus Professor of Social Psychology at The George Washington University, has specialized in psychological factors in the wars of the Twentieth Century. In 1988-89, he was President of the International Society of Political Psychology. His books include *Nobody Wanted War: Misperception in Vietnam and Other Wars* (Garden City, NY: Doubleday, Revised Edition, 1970), *Fearful Warriors: A Psychological Profile of U.S.-Soviet Relations* (New York: The Free Press, 1984). He also edited *Psychology and the Prevention of Nuclear War* (New York: New York University Press, 1986).

Introduction

Burns H. Weston

The inspiration for this book came shortly after its predecessor, *Toward Nuclear Disarmament and Global Security: A Search for Alternatives* (Boulder, CO: Westview, 1984), had gone to press—specifically, on or about March 23, 1983, when President Ronald Reagan, calling on the scientific community "to give us the means of rendering [nuclear] weapons impotent and obsolete," first proposed his Strategic Defense Initiative (SDI). However much, at the same time, the President then may have credited the avoidance of East-West war to the strategy of nuclear deterrence over the preceding three and a half decades, his "Star Wars" proposal implicitly conceded the frailties of that strategy and thus legitimized a renewed search for alternatives to it—a search that, since Senator Joseph McCarthy's time, in the United States at least, had been abandoned by all but the most politically fearless.

President Reagan's distrust of nuclear deterrence as a guarantor of national and international security was—and still is—fully warranted. Nuclear deterrence does not guarantee against human perfidy or error. It does not ward off technological malfunction or mechanical breakdown. It does not ensure that "crazy States" will play by the same rules. It does not prevent paramilitary terrorism or even fairly large-scale conventional wars from affecting perceived vital interests. And, obviously, it is impotent in the face of life-threatening ecological deterioration, explosive economic disparities, and revolutionary nationalism, to name just a few of the *non-strategic* dangers that challenge national and international security worldwide. What is more, in its purely defensive mode, nuclear deterrence leaves all the initiative to the adversary. In its requirement of unambiguous and credible threat, it denies flexibility

1

to political leadership. And in its insistence on invulnerable retaliatory capability in the face of continuous technological advance, it is inherently unstable. Indeed, contributing significantly to economic decay, the erosion of political liberty, and the perversion of basic morality, nuclear deterrence augurs insecurity in the most pervasive sense. All the while, of course, it threatens major extinction.

But the appropriate path to national and international security is not, as President Reagan – and so far apparently President George Bush – would have it, via SDI or some other "technological fix," for reasons well explained in this volume. Scarcely anyone in the scientific community, or anywhere else for that matter, takes seriously the proposition that SDI can provide the means to sweep away the nuclear arms race and the "delicate balance of terror" that accompanies it, much less that it can provide genuine and lasting security for the United States or the world. Indeed, most responsible observers agree that SDI is entirely consistent with – actually supportive of – nuclear deterrence and that if developed and deployed it would repudiate solemn arms control agreements and trigger an unbridled arms race between the United States and the USSR as never before.

Nor is national and international security to be found, over the long term at least, through "minimum deterrence," a strategic policy that presupposes the continued acceptance of nuclear weapons and therefore the ever-present possibility of aggravated annihilation. Those who argue that nuclear deterrence, whatever it has *not* accomplished, has successfully prevented the outbreak of thermonuclear world war for forty-five years adopt a too simplistic calculus of world affairs and ignore the truism that the lack of an effect seldom, if ever, constitutes proof of anything. There of course is no denying that there has been no nuclear war since 1945. But an honest, objective analysis suggests that this fact owes more to economic interdependencies, to the wisdom of foreign leaders we routinely deprecate, and, in many cases, to plain dumb luck than it does to visionary insight on the part of those who have conducted the arms race allegedly on our behalf. Like nuclear deterrence in general, minimum deterrence risks the fateful absence of interdependencies, wisdom, and luck.

In sum, SDI and minimum deterrence, representing traditional approaches to coping with conflict (the first purporting to remove threats to national and international security, the second menacing retribution ostensibly without further endangering national and international security), are grossly inadequate to the task of nurturing and achieving genuine and lasting world peace. Alternatives to these traditional approaches and to the nuclearism that inheres in them thus demand to be invented, developed, evaluated, and applied. The time is long past due to forswear living *with* nuclear weapons. The time is long past due to commit ourselves to living *without* them.

Any alternative to nuclear deterrence, however, must be informed by at least the following three understandings: (1) that conflict is likely to be violently expressed in the world system for years to come; (2) that people want to be protected against international violence and thus are not easily dissuaded from a nuclear deterrence system that *seems* to have worked for better than four decades; and (3) that there must be established, therefore, an *effective* alternative or set of alternatives to nuclear deterrence if we are to escape its mind-boggling risks and secure genuine and lasting world peace. Unless these three understandings inform the search for alternatives to nuclear deterrence as a means of dealing with national and international security, any proposal that might be made will simply prove insufficiently responsive to the real world in which we live, and therefore will be doomed to utopian irrelevancy.

This collection has been assembled with conscious regard for these three understandings and for this reason reflects the belief that genuine and lasting world peace can be achieved only through a complex or medley of interdependent—mostly "soft path"—policy options that *jointly* can define a world free of nuclear weapons and the threat of nuclear war. It reflects the belief that any global security system, if it is to be an *effective* alternative to nuclear deterrence, must be conceived as an integrated plan. Thus, except for the eighth and final chapter, which undertakes to assimilate, appraise, and amend the first seven, all of the chapters that follow have been written from a distinct disciplinary perspective—military, technological, legal, economic, political, psychological, and religious—yet with a common appreciation that no single set of policy initiatives, only a composite of them, can serve as an alternative to nuclear deterrence that is capable of effectively safeguarding core national and international interests. The struggle for a post-nuclear world and the abolition of war more generally is defined by so tangled a web of interdependent and interconnected issues and problems that it cannot be properly comprehended, let alone won, without simultaneously specialized and integrative thought and action.

All of which should serve to remind us that the search for alternatives to nuclear deterrence, and to war itself, must extend beyond the traditional boundaries of political science and military-strategic studies (two disciplines that have faltered in these respects and that bear, therefore, much responsibility for having relegated the search to the underfunded and marginalized disciplines of peace and world order studies).   Major social problems – even minor ones – do not arise along unidisciplinary lines; hence, they cannot be solved by unidisciplinary devices; and the point is only magnified, certainly, when it comes to the problem of achieving a non-nuclear world or, more generally, a world without war.   Of course, a serious commitment to interdisciplinary and transdisciplinary research and policy planning is not an easy one, requiring as it minimally does a radical psychological departure from that "splendid isolation" that nurtures and comforts most theorists, especially those in academe. Nor is it one that sits easily with academic deans and department heads who tend often to perceive a threat to their power and wealth whenever already scarce human and material resources are asked to be shared.   But, surely, significant inroads upon the search for credible alternatives to nuclear deterrence, including the preparation of future citizens and leaders for effective action, are scarcely going to materialize without readjustments of this kind.   A collateral purpose of this collection is to stimulate attention and movement in this regard.

Three last points.   First, the title of this volume, *Alternative Security: Living Without Nuclear Deterrence*, is in a certain sense misleading, implying as it does a rejection of one kind of security for another.   The plain truth is, however, that there is no real security in nuclearism, but, rather, historically unparalleled *in*security – *i.e.*, quite the opposite – hence no security to reject.   Which of course leads one to ask why such terms as "common security" (the Palme Commission's preference) or "comprehensive security" (the Soviet preference) were not employed.   The answer is simple.   Although in no way intended to signify a repudiation of these terms – indeed, all of the ensuing chapters endorse their intent and meaning either explicitly or implicitly – the expression "alternative security" does imply a rejection of a *false* security and therefore carries with it an element of conscious choice.   As stated above, the time is long past due to forswear living with nuclear weapons and to commit ourselves to living without them; the time is long past due that we make that conscious choice.

Second, the time for making the conscious choice is not only long past due but supremely urgent. The revolutionary changes that rocked Eastern Europe and the Soviet Union — indeed, the world! — in 1989-90 and that continue to do so as we hurtle to the end of the Twentieth Century have rendered obsolete the political and military arrangements that assisted European and wider world security for over four decades following World War II. The Soviet Union, home to more than 25,000 nuclear weapons and many newly-awakened nationalisms, faces a world history that demonstrates little support for the proposition that collapsing empires fade quietly. And in our increasingly high-tech world, with military R&D fast at work on atomic guns, particle-beam cannons, and other space-age deviltries that divert attention from the perils of nuclear proliferation, dozens of regimes in the Middle East and elsewhere are acquiring nuclear and other weapons of mass destruction — and the means to deliver them to almost anywhere on earth — with frightening ease and speed. The need for a nuclear-free world is thus mandatory in the most pressing sense, and the more compelling precisely because of these and related hazards.

Finally, the time for making the conscious choice — to reject nuclear deterrence in favor of a set of credible and effective alternatives to it — is consummately ripe just as it is long past due and urgent. For the first time in more than forty years, not least because of *glasnost* and *perestroika*, serious interest in procedural and structural change is reviving. Talk of transforming the North Atlantic Treaty Organization (NATO) and the Warsaw Treaty Organization (WTO) from predominantly military to predominantly political cooperatives, for example, is now a matter of increasingly routine discourse in the professional journals and the popular media, pointing to the invention and construction of a new European security order. And there is, above all, speculation as to the future integration of Western Europe as of 1992, and of possible all-European unity in the foreseeable — perhaps even near — future. Indeed, it is not now far-fetched to wonder aloud whether Europe, which gave birth to the modern State system, may not provide the crucible of interests and values that will bear witness to that system's transcendence. In short, though each author in this volume was asked initially to suspend his present sense of contemporary reality sufficiently to imagine a world more or less free of nuclear weapons (for the purpose of ensuring recommendations that could serve as visions for, as well as programs of, change), the events of the last years of the 1980s, especially the profound and profoundly exhilarating events that began to take place in the Soviet Union and Eastern Europe, lead one to appreciate that "reality" is never fixed and that the magnitude of the struggle for a world free of nuclear weapons is not so overwhelming that it is beyond human capacity.

Hence these chapters. They are tendered in the hope that political leaders and citizens everywhere will take seriously the fact that there *are* realistic alternatives to the horrible undertaking in which they daily participate and that it is not so much for lack of ideas than for lack of political vision and will that we fail to move toward them rapidly.

1

The Military and Alternative Security: New "Missions" for Stable Conventional Security

Thomas F. Lynch, III

It is the soldier that knows all too well that it is best that war is so terrible; else we would grow too fond of it.
— General Robert E. Lee, 1862

"[M]odern weapons," writes Steven Meyer, "make it impossible for states to defend themselves by military-technical means alone. War means catastrophe. . . . The military's intimate involvement in arms control has thus become a reasonable proposition in Soviet eyes."[1] A mere five years ago, such a statement about the Soviet Union would have been considered inconceivable. As the most militaristic of the major world powers, the USSR was perceived universally as a captive of the view that military power is fundamental to State prestige; and, accordingly, popular wisdom had it that the Soviets would remain committed to the acquisition of nuclear and non-nuclear arms. This belief relative to Soviet intentions contributed, of course, to a spiral model of US-USSR military relations, in which Washington vowed to match Moscow missile for missile, tank for tank, and *vice versa*. And, taking their cue from the superpowers, the rest of the world seemed equally wedded to a security paradigm that features (1) States acting out of pure self-interest, (2) a worldwide proliferation of conventional arms, and (3) a general consensus that security rests on the maintenance of nuclear deterrence. Against this backdrop, appeals for denuclearization and demilitarization by transnational actors such as Greenpeace and international assemblies such as the Pugwash Conference appeared bleak. Prospects for actual movement toward newer, less ominous security regimes appeared bleaker still.

In the intervening years, however, traditional thinking has been challenged by new realities. Formerly bleak prospects now appear as promising opportunities. To a degree believed unimaginable until just recently, the prospects for a cooperative international security regime transcending the post-World War II straitjacket of strategic nuclear deterrence seems a genuine possibility.

This development parallels, of course, the political and intellectual ascendence of Mikhail Gorbachev,[2] the first leader of a major world power to argue that "collective security" is today the only way for nations to assure themselves real security. As Gorbachev observed in *Perestroika*, "the human race has entered a stage where we are all dependent on each other. No country or nation should be regarded in total separation from another, let alone pitted against another."[3]

Conspicuously signaling this dramatic shift in security thinking was, of course, the Intermediate Nuclear Forces (INF) Treaty between the two superpowers in 1987, eliminating an entire class of nuclear missiles in the form of "intermediate-range and shorter-range missiles."[4] Another indication was the ongoing Conventional Forces in Europe (CFE) negotiations in Vienna, which sought to eliminate, for the first time in history, large numbers of the superpowers' conventional military arsenals, and probable "deep cuts" in their strategic nuclear arsenals as well, via the Strategic Arms Reduction Talks (START) scheduled for 1990.[5]

All of which cannot help but fuel speculation that the international security regime the world has known since 1945 is due for — indeed, is already undergoing — basic change. Security based on the nuclear balance of terror will be hard pressed to survive in the coming years as emerging economic and environmental concerns take precedence over the "logic" of Mutual Assured Destruction (MAD).

It is my contention that these and related historic developments make a world security regime that is not dependent on nuclear deterrence a real possibility in the near future.[6] I take pains to make this point at the outset because, as a military professional, I am well aware that the accepted — and "consumable" — military view continues to be that nuclear weapons can and *must* remain the foundation of any realistic international security regime. Certain proponents of the US Strategic Defense Initiative (SDI) make just such a claim in their advocacy of that controversial program.[7]

It is my considered belief, however, that the momentum toward denuclearization is proceeding at a pace and a logic that leaves such "old thinking" seriously out of touch with reality; and, moreover, that it is possible, even in military circles, to contemplate a *realistic* global security system in which the world will be sufficiently free of nuclear weapons to eliminate reliance upon them as a matter of routine. Like it or not, the forces of history are placing the militaries of the various sovereign States at a major crossroads, obliging them, and the political constituencies that control them, to face up to the fundamental question of what posture they should assume—what role they should play—in a security regime wherein nuclear weapons are no longer dominant.

But there is an important caveat: If my contention is to prove ultimately correct, the national militaries must be allowed *and encouraged* to participate actively in the evolution of alternatives to nuclear deterrence, emphasizing their reduced warfighting and enhanced crisis-prevention capabilities; and I hasten to add that, though a penchant for conservatism tends to be in the nature of the military beast, as Samuel Huntington observed some thirty years ago,[8] this proposition is no idle one. Popular impressions to the contrary, leading thinkers in military circles increasingly are giving serious thought to such "progressive" options as crisis prevention, conventional arms control verification, military information exchanges, and coordinated force restructuring—clearly signaling the possibility if not yet the actual promise of a global security regime less homogenous than the nuclear threat system that has prevailed since 1945. Despite long-standing resistance to any alteration of the postwar security order, the world's militaries are today showing real signs of genuine (and politically motivated) support for a global security system that features at least improved military cooperation and reduced risk of destructive war without a radical restructuring of the common notions of security.

In sum, assuming an active definitional and implementing role for the world's national militaries, I am persuaded that a viable alternative to nuclear deterrence is within reach in the near term—specifically, a modified, military-enforced variant of conventional "non-offensive defense" (or "defensive defense") by the year 2000. I believe also, again assuming an active definitional and implementing role for the world's militaries—and assuming, too, the appropriate *political* will—that even more progressive non-nuclear security regimes are possible in the foreseeable future. Before attempting to assess the strength of these possibilities, however, it is helpful to have in mind the breadth of options, both "conservative" and "progressive," that are at least theoretically at hand.

THE ALTERNATIVE MILITARY SECURITY CONTINUUM

There exist numerous alternative options to the current nuclear-dominated security paradigm. If we look here only at the most prominent, they can be subdivided into *conservative* and *progressive* categories—the former tending to be more commonly associated with the military repertoire of the past; the latter tending, at least formerly, to be spurned by the military altogether. A quick review of their underpinnings reveals the logic of their subdivision.

Conservative Options

Conventional Deterrence (Arms Race)

The most common of the conservative options, with us since before World War II, can be labeled "Conventional Deterrence Without Bounds" (CDWB) or, less flatteringly, an unrestrained conventional arms race. Its features are familiar. In an essentially post-nuclear world still lacking in centralized institutions capable of managing world order, independent sovereign States would build conventional weapons as they would see fit, ostensibly maximizing their security by stockpiling technologically sophisticated conventional weapons in sufficient quantities to hold potential adversaries at bay. These weapons would have the additional advantage of holding at risk the territorial sovereignty and diplomatic maneuverability of States lacking equivalent stockpiles.

Simple and blunt, this option remains appealing to those of a decidedly conservative bent. But a non-nuclear alternative security regime based upon CDWB has decidedly unattractive features as well, sufficient to make it an unrealistic alternative, especially in the present international political climate.

In the first place, it promises to be extremely expensive for participating States. In fact, this was precisely the concern of the Truman Administration when it opted for the unilateral development of the atomic—and then hydrogen—bomb in the face of an ominous Soviet conventional threat; and it was the concern, too, of the Eisenhower Administration, inspiring John Foster Dulles' famous sojourn into the world of "massive retaliation."[9] The theory of the postwar American political establishment was that the US comparative advantage did not lie in a quantitative arms race with the more determined Soviet Union. Thus, a cheaper alternative was seen in the exploitation of a then asymmetrical advantage in the nuclear realm. Now, forty years later, the fact that the nuclear option has become an expensive proposition in its own right tends to obscure the fact that an unrestrained conventional arms race would guarantee the United States, now as in the 1940s and 1950s, great cost and little in the way of enhanced security. Indeed, it is precisely for these reasons that the Soviet Union is at present strongly supporting drastic military cuts both in its own military forces and in those of its Warsaw Pact allies; and it is for these reasons, too, that, as I write, economic and political discourse in the United States is swinging markedly in the direction of as much as a 50 percent reduction in defense spending over the next five to ten years.

CDWB promises little in the way of enhanced security for the rest of the world's States as well. The majority of the world's smaller States depend on the larger ones for conventional arms in the first place; less well endowed economically, they have little ability to develop the economies of scale that are needed to procure modern conventional weapons, and thus remain dependent on the military giants. More problematic, the giants would have less incentive to make sophisticated conventional weapons available to smaller States if a lead in conventional weapons technology would serve as the new underpinning for absolute military power. Large States would have great incentive to hawk only technologically inferior weapons to the small States, and thus the relative technology differential of the present security order would remain unaddressed, tending toward a self-fulfilling steady-state condition.

Finally, CDWB offers little prospect for a meaningful reduction of damage should a major conflict break out. The massive destruction wrought by conventional weapons during World War II serves as abundant testimony. The destructiveness of conventional weapons has only "improved" since 1945 – making even the use of the term "conventional" misleading – so that the marginal difference in damage between an all-out nuclear war and an all-out conventional war might be so small as to make arguments over which is preferred the worst form of pedantry. Hence Mikhail Gorbachev's observation concerning the continent of Europe, which has significance for the world at large: Not only are conventional weapons many more times destructive than they were in World War II, but the large number of nuclear power plants and major chemical works that would likely be destroyed in a large conventional conflict would make most of a postwar world uninhabitable.[10]

Strategic Defense Initiative

Until fairly recently, a popular second conservative alternative to nuclear deterrence was the Strategic Defense Initiative, first made public by President Ronald Reagan in a nationally televised speech on March 23, 1983.[11] As explained in the President's speech, SDI (or "Star Wars," as it came to be known) was posed initially as an "impenetrable shield" against incoming ballistic nuclear weapons and later—and more significantly for present purposes—as capable of rendering nuclear weapons useless.[12] Of course, the promise of Star Wars would be a monumental breakthrough in advancing a defensively based alternative security system if such grand claims were in fact attainable.   Nuclear weapons would be rendered obsolete by an impregnable shield and reliance upon them would come to an end.

Six years of acrimonious debate and many precious dollars later, however, the military (like the public at large) is well aware that SDI is nowhere near the panacea it was initially billed to be. The technology itself has proven extremely problematic.[13] Scientific experiments indicate that the potential for a laser-generated, ground-based anti-missile interception shield is marginal for the foreseeable (and perhaps permanent) future. Land-based projectile interceptors are manifestly vulnerable to countermeasures, ruses, and saturation; and space-based systems are susceptible to each of these foils, plus being vulnerable to "killer satellites." Indeed, the whole prospect for space deployment of SDI would serve only to militarize outer space in clear contravention of previously negotiated bilateral agreements, without assuring any tangible upgrade in security.[14]

But perhaps even more important than the technical problems associated with SDI is the fact that the Star Wars program threatens the foundations of reciprocal restraint in those areas of the nuclear arms race that are likely to produce instability and a spiral escalation in arms race competition. The provisions of the 1972 Anti-ballistic Missile (ABM) Treaty[15] recognize such an offensive-defensive arms race spiral as an inherent risk of anti-missile technology, and the operational review precipitated by the Star Wars debate has only reinforced the wisdom of this position. SDI would likely provoke rather than deter preemptive strikes. In addition, SDI would prove extremely costly.

Thus, SDI appears to represent a giant step backward to the bleak security arrangements of the mid-1960s, not a step forward to an improved non-nuclear international security regime. The weight of technological, economic, and legal drawbacks has made SDI suspect as a truly viable security alternative; and, to their credit, the Bush Administration and the Congress appear to be coming around to its illogic as of this writing, albeit slowly.[16] It is and should remain a stillborn concept, anachronistic in the changed political climate of growing East-West understanding and cooperation.

Progressive Options

Three progressive alternatives to nuclear deterrence that have received the most attention are (1) non-offensive defense, (2) civilian-based defense, and (3) an independent international peacekeeping force. The first especially is the focus of increased military attention because of the manifest recent change in the international security climate and because of the increasingly apparent infeasibility of the conservative alternatives noted above.

Non-Offensive Defense

The first alternative is non-offensive defense (NOD). Though recently popularized by Soviet President Gorbachev in terms of the buzzword "defensive defense,"[17] the intellectual underpinnings of this alternative to nuclear deterrence are found in the thinking of Western Europeans. Inspired by fears of destabilization and destruction to Europe because of NATO's evolving strategy of "flexible response," theorists such as Anders Boserup and Horst Ahfeldt pioneered thinking about how the world's militaries (particularly in Europe) might reduce or eliminate the threatening offensive capabilities of their armed forces while retaining a viable, non-threatening defensive posture with those forces at the same time.[18]

NOD has numerous incarnations, too many to elaborate fully in this chapter.[19] However, each of these incarnations adheres to a common set of principles, the most salient of which for a post-nuclear world include the following:[20]

1. orienting weapons procurement away from mobile, long-range systems (deemed more threatening and, thus, destabilizing) toward defensive anti-aircraft and anti-armor systems;
2. moving conventional force strategies away from unbounded mobility and counteroffensive capabilities toward those featuring a wide array of *fortified* defenses with *some* counterattack capability (at the tactical rather than the operational or strategic level) retained to deny the aggressor tactical advantage and breakthrough; and,
3. decentralizing and dispersing defensive forces to reduce the opportunity for the offensive side to target and disable the defense effectively (with proponents of this course of action emphasizing the value of new communications technology in keeping such a force in contact for coordinated defensive action despite the extensive dispersion of forces).

The most recent advocates of NOD have included, also, calls for its implementation simultaneously with civilian-based defense (*see* this chapter: *Civilian-Based Defense*).

In essence, NOD has two principal constructs. Each features a move toward depth and density of the battlefield.

The first construct, referred to as the "Spider in the Web" or the "Study Group on Alternative Security Planning (SAS) plan," is closely linked to Horst Ahfeldt and the West German Social Democratic Party (SPD). It has four essential elements:

1. *"non-provocative" light-infantry commandos* with modern anti-tank weapons, assisted by an elaborate network of static defensive barriers and fortifications forming the forward defense (the "web" in the SAS plan);
2. *armored forces*, smaller than those at present and redeployed well behind the "web" to serve as "spiders," capable of rushing to the point of an enemy breakthrough to drive the attackers back out of the web, but incapable of continuing in an effort to conduct deep attacks into enemy territory;
3. *artillery*, used to back up both the commandos and the "spider" forces, and eventually to be phased in to replace the armor units altogether; and
4. *sophisticated communications*, used to tie the entire network together.

This type of an alternative security regime has been proposed for unilateral implementation in the West by both Ahfeldt and the SAS.[21]

A second construct features a more qualitatively based effort to reduce the conventional arsenals of major belligerents in a reciprocal manner. Its principal model was embodied in the 1987 Jaruzelski Plan for Europe, named for General Wojciech Jaruzelski, current President of the Polish Council of State, who first proposed it.[22] It features three principle elements:

1. conventional weapons to be gradually reduced starting with those with the strongest capability for surprise attack;
2. military doctrines to be modified, on a multilateral basis, so as to conform more strongly with strictly defensive goals; and
3. military-to-military and military-to-civilian confidence and security-building measures to be gradually multiplied and strengthened, with an eye toward making verification of conventional defensive postures a reality before the next century.[23]

It is noteworthy that Mikhail Gorbachev's pathbreaking address to the forty-third session of the United Nations General Assembly on December 7, 1988, closely mirrored these aspirations and contributed a Soviet overture to get the "ball rolling" toward the Jaruzelski Plan (albeit initially in a unilateral "restructuring" manner) before the end of 1990.[24]

Civilian-Based Defense

A more passive alternative to nuclear deterrence than NOD is civilian-based defense (CBD), which, in its purest form, postulates the elimination of all formal military forces and a reversion to society-wide, non-violent resistance of any hostile invasion. The idea is that, properly organized, a civilian-led, collective resistance can effectively deny an aggressor the benefits of conquest. As one of its chief proponents, US sociologist Gene Sharp, has put it, CBD would threaten an aggressor with a "non-violent blitzkrieg," *i.e.*, near total non-cooperation by the invaded populace such as would bewilder the attacker with society-wide paralysis, a paralysis that would deny the attacker the "fruits of victory" and thereby serve as an effective deterrent to invasion.[25]

But what if such deterrence should fail? According to Sharp, the invaded country would conduct its resistance through a series of non-violent tactics designed to make the occupied country ungovernable and thus "undigestible": (1) *conversion* of the adversary toward the societal values of the conquered State, a tactic featuring parades, marches, vigils, and the like; (2) *accommodation* of the invaded populace to the goals and objectives, in modified form, of the invader via fasting, competing news dissemination, and establishment of parallel government structures; (3) *non-violent coercion* of the administrative entity of the invading force toward the rules and regulations of the invaded society, largely through strikes, boycotts, non-violent facility occupation, and other selective acts of sabotage; and (4) the *securing of international sympathy and support* via the foregoing actions to pressure the invaders into abandoning their occupation.[26] Faced with the prospect of such tactics, the theory goes, a potential aggressor would think twice before invading. Similar conclusions are drawn by the British-based Alternative Defence Commission.[27]

Recognizing, however, that CBD would not be easy to implement even under the best of circumstances because it would require education and planning on a grand scale, Sharp in particular advocates a transitional phase from current security to CBD, which he calls "transarmament."[28] In this phase, the entire populace would need to be informed and educated in the principles, purposes, and procedures of CBD, and military forces would have to be trained to conduct passive, resistance-based defense as a supplement to weapons-based security structures. Thereafter, once confidence in CBD were to grow, CBD itself would come to supplant weapons-based defense as a whole. To his credit, Sharp acknowledges the ambiguity inherent in this transitional phase, and that it would be a most difficult undertaking.

An Independent International Peacekeeping Force

The last and possibly most progressive alternative to nuclear deterrence as a means of ensuring national and international security is that which features an independent international peacekeeping force as the ultimate guarantor of national territorial sovereignty and world peace. Among the most comprehensive and systematic of all such proposals, said to represent "the classic formulation of the world federalist structure of government,"[29] is the so-called Clark-Sohn plan, *World Peace Through World Law*.[30] First published in

1958 and subsequently extensively revised, it calls for a mandatory global security system in which, among other things, a truly global armed force—a United Nations Peace Force—would alone serve as the effective "policeman" of the world. To come into being after the elimination of all national military forces and to have among its principal purposes the enforcement of decisions rendered by a supranational adjudicative authority committed to the world rule of law, this UN Peace Force would be limited to a full-time standing force of 200,000 to 600,000 members, with a Reserve of 600,000 to 1,200,000. It would include a body of conscripts "individually recruited" without regard to their national affiliations and limited in service time to a period not to exceed ten years. Also, to encourage competent participation, it would feature a number of incentives (pay and retirement, among others) necessary for its credible establishment and maintenance.[31] Thus, sophisticated and detailed, the Clark-Sohn plan is a comprehensive attempt to make an essentially anarchical world more hierarchical in its maintenance and use of force. It recognizes that an essentially disarmed world cannot prevent unilateral resort to force in the absence of a credible and superior military watchdog capable of ensuring that national military forces do not reemerge in disregard of community-wide prohibitions to the contrary.[32]

RECENT PROGRESSIVE MILITARY UNDERTAKINGS

How does one evaluate the prospects for these military and quasi-military options for a post-nuclear international security system? As appealing as some of them may be theoretically, what are their chances in the real world of military thinking and planning?

A useful place to begin is with the trends that appear now to be evident *and increasingly stable* in the military dialogue that has taken place between the two superpowers in the last few years. As the US-USSR security dialogue goes, it may be said, so go the prospects for the rest of the world[33]—at least to a large extent.

At the US-USSR summit in December 1987, at the time of the signing of the INF Treaty,[34] the defense ministers and heads of the armed forces of these two traditional adversaries agreed to a timetable for increased military visits over the ensuing four years.[35] Also, they agreed to a specific agenda of cooperation in crisis prevention, arms control verification, and information exchange[36]— each central concerns of Admiral William J. Crowe, then head of the US Joint Chiefs of Staff (JCS), and Marshall Sergei F. Akhromeyev, then Chief of Staff of the Soviet military. All of which set into motion a dramatic increase in US-USSR inter-military dialogue and expanded military participation in the shaping of arms control/arms proliferation and crisis prevention policy, including, in a highly patterned manner, an unprecedented number of US-Soviet exchanges of military personnel and ideas. But what truly is significant about the December 1987 accords is that they denoted that the heads of the two most resource-rich militaries of the postwar era had come to realize that funds previously committed to costly military hardware for the purpose of maintaining warfighting capabilities now were going to be diverted to less costly, alternative means of ensuring national security. They denoted, too, that the respective military "chiefs" had come to realize that they had to buy in to a "softer" path of ensuring national and international security if they were to have any control whatsoever over the fiscal resources that were likely to be diverted from their professional jurisdictions.

The December 1987 agreement and its aftermath produced both acclaim and acrimony. The unprecedented 1988 visit of Soviet Chief of Staff Akhromeyev to US military bases and exercises, and the equally pioneering visits of Secretary of Defense Frank C. Carlucci in late 1988 and JCS Chief Crowe in June 1989 to counterpart facilities in the Soviet Union won universal acclaim for advancing understanding between the two most opposed military establishments of the postwar world. But parallel initiatives between the two militaries have not been as popular. In particular, the crisis prevention protocol agreed to between the two militaries[37] and some of the more aggressive military efforts to find common ground for technical, doctrinal, and strategic exchanges have been greeted with rancor in civilian circles. Both the US Department of State and the Soviet Foreign Ministry have expressed concern that the growing involvement of the military in the *diplomatic* realm is a direct threat to traditional civilian ascendancy in these areas.[38]

Sensitive to these expressed concerns, the US military, with a vision heretofore unseen in the postwar era, has nevertheless pressed cautiously forward, determined to expand its participation in the security fields of crisis prevention, arms control verification, military information exchange, and force restructuring, none of which have been the focus of much military interest or concern traditionally.[39] The manner in which the US military has pursued each of these fields demonstrates the extent to which the military bureaucracy feels comfortable in aligning itself with policies that many say are essential for a progressive post-nuclear international security system.

Crisis Prevention

The most likely scenario for the outbreak of war between the two superpowers is one in which their respective forces come into accidental contact, resulting in low-level incidents spiraling into a holocaust. Accordingly, the US military has recently moved aggressively to establish ground rules with the USSR on the procedures that should be followed if and when their respective military forces come into accidental contact. Stimulated by the 1983 KAL 007 incident, by instances when US and Soviet forces have in fact come into close proximity,[40] and by President Gorbachev's commonsense argument that even a small, inadvertent conventional war in Europe would likely turn into a global nuclear war because of the density of nuclear power plants in the region,[41] the leaders of the two military establishments signed, in early June 1989, a joint protocol calling for a regularized crisis prevention procedure to be followed by the military forces of the two countries if and when they come into contact.[42] Although a precedent for this type of communication has existed since the first significant US-USSR nuclear arms control treaties of 1972, it was not until the protocol of 1989 that talks concerning crisis prevention at lower levels of US-USSR military contact were taken especially seriously.[43]

At the same time, the military chiefs of the two superpowers have seen to the development of a military-to-military working group to continue the evolution of crisis prevention rules and procedures. Created in 1988, the working group is chartered to accomplish a variety of tasks that are designed to allow the military forces of the two countries to anticipate, mediate, and resolve the conflicts that ensue when the two nations' military forces – sea, air, or land – are brought into close proximity.[44] Among these tasks are the study of the doctrinal procedures and tactics of the adversary's forces (in an effort to distinguish hostile actions from routine defensive procedure) and the formulation of communication and distress channels designed to assist military elements in close proximity to decipher and defuse confusing and/or threatening situations.[45]

In addition, the US and Soviet militaries have been pushing initiatives designed to build confidence between the two countries,[46] including the curbing of the proliferation of illegal arms transfers resulting from both unintentional military losses of equipment and deliberate larceny due to lax security and accountability procedures.[47] Also, procedures for the identification and interdiction of illegal drug flows are now being shared, with an eye toward making multinational progress in an area where individual sovereign efforts appear doomed to fall short and only fuel acrimony between parties of diverse means and objectives.

In sum, new US-USSR military cooperation in the area of crisis prevention has been expanding, and this bodes well for the search for an alternative to nuclear deterrence as a means of safeguarding core national and international interests. The cooperative spirit being established between the US and Soviet militaries on issues concerning common incidental contact can assist the viability and eventual stability of post-nuclear security regimes that will require procedural ties and better understanding between independent, sovereign defense forces. These crisis prevention initiatives do not actively anticipate a policy as radical as that of eliminating sovereign national military forces in favor of some independent international peacekeeping force; but they do facilitate the information exchange that obviously will be necessary to make any viable non-nuclear security regime a reality.

Arms Control Verification Means

Since 1988 there has been a mini-revolution in the degree to which the superpowers have agreed to forms of arms control verification beyond those strictly defined by national technical means. Indeed, in many formerly taboo areas the civilian-military establishments of the two countries have agreed both in principle and in practice to allow substantial intrusive forms of *human* intelligence-gathering as the basis for arms control verification.

This trend has been pioneered in the nuclear weapons arena. Beginning with the INF Treaty[48] and the attendant bilateral observer parties established in conjunction with the Nuclear Risk Reduction Centers set up in Moscow and Washington in 1987-88, intrusive on-site verification that could only have been dreamed of in the early 1980s has become a reality of arms control implementation in the early 1990s. And it has paved the way for parallel developments in the military-dominated conventional weapons arena. As already noted, civilian oversight of arms control initiatives in the arena of conventional weapons is much more problematic than it is relative to nuclear weapons, due both to the number of systems involved and to their special technical nature.[49] As a result, the US military has been quick to capitalize on its comparative technical advantage to establish its verification prowess in this new realm of perceived opportunity.

With respect to monitoring and reporting on adversary training exercises, for example, the military can point to long-standing experience and expertise sufficient to warrant a continued and expanded role relative to verification. Ever since the Huebner-Malin Agreement of 1947,[50] the United States and the Soviet Union have had joint observation missions established at each other's European defensive alliance headquarters – the Soviet Military Liaison Mission (SMLM), jointly located with the US forces in Frankfurt, and the American Military Liaison Mission (AMLM), jointly located with its Soviet counterpart in Potsdam. Through 1987, these two missions existed as small and eclectic legalized "spy" headquarters. Their military members, in specially marked and plated vehicles, were allowed to serve as observers of military exercises of a large scale and to travel to other areas of the rival alliance's countryside albeit with special restrictions to prevent them from getting too close to major installations, training sights, or planned defensive positions. Recent bilateral agreements emphasizing openness and confidence-building have enabled the respective militaries to expand their conventional force observations to the entirety of the opposition's military activities.

The Stockholm Conference of 1984 set the stage for greatly expanded conventional exercise observation rights by opposing military observation teams.[51] This initiative was codified by verbal agreements reached between President Reagan and General Secretary Gorbachev (over the muted protests of military advisors) at Reykjavik in 1986 and later formalized in a multilateral protocol at Stockholm, also in 1986,[52] calling for greatly expanded notification and an increased number of invitations to observe the standard military exercises and training events of the rival bloc's military activities. The joint protocol also enabled the observer military forces to include more personnel than those previously detailed for duties on the official military observer force. For the US-NATO side, this meant having to give at least two weeks' notice to the Soviets of any exercise or training event involving 20,000 or more NATO soldiers and, in addition, to facilitate the *complete* observation of these exercises by Soviet/Warsaw Pact line officers hastily assembled from regular military units just across the East-West border. In a significant departure from past procedure, this opened "window" of observable training now allowed Soviet military observers to inspect personally all of the NATO forces and training areas in the geographic location of the announced exercise whether these forces and training areas were directly in the exercise events or not. These conditions were to be true, beginning in late 1987, for the Soviet-WTO side as well. Thus, by the end of 1987, the US military found itself swallowing hard as the terms of the Stockholm observation agreement facilitated the heretofore unthinkable: Soviet generals and colonels known to be in charge of regular Warsaw Pact ground force structures allowed to roam freely over the vegetation-bare firing ranges of the largest NATO training complex in West Germany, Grafenwoehr; and to climb atop and witness, *in person*, the previously well-classified battle-drill of US M1 tanks working with their supporting Apache helicopters and A-10 attack aircraft! The same previously unfathomable scene was soon thereafter played out on the Soviet-WTO side as well. The military cloak of secrecy had truly been thrown off in favor of *glasnost*.

This opening of what many lower-ranking military officials feared was a Trojan horse actually was a calculated bid by the military hierarchy to "go with the flow" and stake a claim as the rightful inheritors of the resources necessary to build the complex organizations required for verification of the conventional arms control foreseen for the coming decade. Featuring movement in the direction of more widely observed training events and more open discussion concerning tactics and doctrine, the military establishment appears to have succeeded in this venture. Currently there is talk in the US military of expanding its observation force both in the US military itself and via a parallel organization within NATO.[53]

Less clear is whether the United States and the Soviet Union will agree to their military counterparts having access for verification purposes to the military hardware industrial production quotas of the other side. This issue has remained a point of contention in the nuclear weapons realm. Hopes for a revolutionary opening of conventional arms production and storage facilities to reciprocal verification by military or civilian-military teams were set back in Summer 1989 during the course of negotiations on chemical weapons; it appears likely that an agreement in the conventional weapons realm will not fare well either. In the Summer 1989 chemical weapons negotiations, both Moscow and Washington expressed grave misgivings over proposals to have their respective chemical industries subjected to intrusive verification from the other side.[54] It is not unreasonable to conclude, therefore, that the same wary feeling will continue in the immediately foreseeable future relative to the conventional arms industry as a whole.

Nonetheless, in a process that has proved painful for military traditionalists on both sides of the East-West divide, the military hierarchies of the Soviet Union and the United States have established themselves in the last few years as the legitimate heirs to the resources needed for conventional arms control verification; and in so doing they have made it reasonable to think that someday they might buy in to relatively progressive alternatives to nuclear deterrence, including some form of NOD, possibly in the coming decade. Of fundamental importance, however, is that the military hierarchies' position as verification "experts" be safeguarded in whatever alternative to nuclear deterrence might ultimately be worked out.

Military Information Exchange

Among allied military nations, historically, there is a long tradition of information-sharing on a whole spectrum of issues ranging from protocol to sophisticated weaponry. Between hostile governments, such information is secured ordinarily via espionage and risk of treason (except in the case of access to information about defense spending in essentially open societies, made readily available by the public debate over such matters that typically finds its way into the free press). Despite this historical pattern, however, US-USSR relations during 1988-90 altered the process of military information exchange between the two superpowers.

In the first place, the number of cooperative ventures between the professional military of the United States and the Soviet Union has expanded greatly since the Akhromeyev-Crowe protocol of December 1987.[55] In 1989, military workshops on strategy, tactics, crisis prevention, and international peacekeeping procedures were held both in Moscow and in Washington, and formal discussions were underway to facilitate exchanges of professional military officers at all levels of the professional education system—from the collegial military academies to the professional staff schools. It is entirely possible that, in the foreseeable future, Soviet military instructors will teach for a year in the United States and that US military instructors will do the same in the Soviet Union. This development is not merely a superficial matter of reciprocal professional courtesy; it has the potential to enhance crisis prevention measurably in the long run, a potential that would be significantly strengthened were such exchanges expanded to embrace all the world's militaries. After all, it was US-educated and trained Japanese Admiral Isoroku Yamamoto who, in the late 1930s, spoke out most strongly against the Japanese government taking up arms against the United States.[56] If there had been a few more cross-culturally educated Japanese officers, Admiral Yamamoto's warnings might have been taken seriously.

Second, the US and USSR militaries have already made good progress in announcing planned field exercises to the other in a manner timely enough to allow observer teams to be present. Combined with national technical verification means to monitor arms control violations, this process has also had the inevitable effect both of promoting "crisis stability" and of building mutual confidence in the adversary's inability and unwillingness to launch a surprise attack. Continued bilateral progress between the Soviet Union and the United States in this realm, too, could hasten multilateral initiatives along these lines and, as a consequence, enhance global security overall.

Finally, progress in information exchange between the Soviet Union and the United States is being made even with respect to the highest order of military information. Recent notes from the July 1989 plenary session of the Conference on Security and Cooperation in Europe (CSCE) relative to procedures for military information-sharing indicate that the Soviet Union and its Warsaw Pact allies may be very close to an agreement in principle regarding an annual exchange of information on domestic conventional weapons procurement and deployment programs,[57] and very soon, both the US and the USSR thus may know how the other plans to spend its defense dollars during a coming fiscal year. Although this agreement of itself may not produce major breakthroughs if some acceptable form of intra-industry verification measures is not concurrently agreed to, the fact that it has been made at all suggests that even more information exchange about highly sensitive matters (and commensurate reduced emphasis on "worse case" guessing about the other side's budget and priorities) will be possible. The US military would be put in a position of greater surety and less risk, and the excesses resulting from past rancorous budgeting procedures could come finally to an end—thereby aiding in the creation of a less destabilizing arms production environment.

Of course, all this progress in military information exchange does not guarantee the development of the confidence-building security alternatives that can result from such exchange. However, it does enhance the prospects for greater mutual understanding between the two superpower military establishments; and this consequence, in turn, can enable the military bureaucracy, all else being equal, to accept more readily the pursuit of those alternative security procedures and structures that bank on the willingness of the world's militaries to work toward a more stable, *ergo* more peaceful, world through enhanced personnel and information exchange.

Changes in Force Composition and Structure

Progressive changes in force composition and structure, which were long believed in military circles to be totally non-negotiable, appear now to be a real possibility in the coming decades.[58] President George Bush's proposal to slash US forces in Western Europe by some 30,000 to a level of 275,000 if the Soviets do the same highlights movement in what heretofore has been a highly circumscribed policy option.[59] Whether this movement ever will develop to the point of a non-offensive defense security option is of course debatable, but the rhetoric prevalent among prominent military thinkers as of 1989, both in and out of the Department of Defense, suggests a greater receptivity to the concept of NOD than there ever has been in US military history.

To a great extent, fiscal reality—an overall faltering economy in the Soviet Union and the balance-of-payments crisis and "budget crunch" in the United States—appears to be compelling less expensive forms of conventional military defense; and given that "qualitative disarmament" palpably contributes to defense budget savings there naturally is political and military interest in this option. For Mikhail Gorbachev, NOD is a way to economic restructuring through military *perestroika*.[60] For the US, it is seen increasingly, among prudent military advisors and planners at least, as a way of defining a politically marketable defense policy in the face of a dramatically lessened Warsaw Pact threat before one is forced upon the Pentagon as a "must-do" proposition by some other agency of government.

What is emerging in the US military, in other words, even in the face of some important reservations about NOD,[61] is a willingness to endure funding cutbacks and doctrinal adjustments in those weapons systems that, on their face, have *relatively greater* offensive potential. Former four-star Army general and presidential advisor Andrew J. Goodpaster noted in early 1989 that, although fraught with potential dangers as well as opportunities, a fundamental restructuring of NATO's forces needed to be studied and undertaken in a prudent manner as soon as possible.[62] The Atlantic Council, which Goodpaster chairs, not surprisingly adopted much the same position in Summer 1989.[63] But especially enlightening is an article co-authored by former Army Chief of Staff Edward C. Meyer in the Summer 1989 issue of *Foreign Affairs*, discussing much the same sort of NATO restructuring as General Goodpaster and the Atlantic Council appear to have in mind and succinctly summing up the cautious yet progressive mood present in military circles on the topic of non-offensive defense:

> If an agreement can be reached with the Soviet Union to move toward forces that pose less of a threat of sudden large-scale attack, NATO could begin to design forces for deployment close to the inter-German border that would be more defensive in character. Such force might include greatly improved intelligence gathering means, secure communications equipment, survivable mobility for soldiers, unattended aerial and land vehicles, surface-to-air defense systems, longer-range precision-guided munitions and other high technology contributions to defense. . . . There will always be a need to move survivable direct firepower in response to enemy armored and other forces. If this function can be performed by means other than heavy tanks, the nature of NATO forces could be significantly altered. Significant research and development funds devoted to this problem, before embarking on a new and heavier tank project for NATO, would be money well-spent on both sides of the Atlantic.[64]

Meyer's bottom line above appears to be *the* bottom line for the US military on the subject of non-offensive defense at present. NOD is acceptable as an alternative to the present situation only if:

1. it is not undertaken unilaterally by the US; the USSR must lead the way and make painful and irreversible cuts to its offensive structure first;
2. it is undertaken as a strategy that does not threaten any traditional military warfighting roles outside of that involving "deep penetration" into enemy territory;[65] and
3. it is accompanied by a monetary commitment to its execution commensurate with the current military budget allocation in the US (in essence, an approach that would make desired military budget savings more difficult to achieve).

These preconditions are, of course, difficult to reconcile with a pure form of NOD or defensive defense as proposed by Mikhail Gorbachev. They also do not align well with the complete elimination of entire classes of weapons or weapons systems called for in the writings of Boserup, Ahfeldt, and others already cited on the subject.[66] They do, however, reflect significant movement in the direction of the NOD thinking that has not been prevalent among Western European analysts for many years.

A MILITARY ASSESSMENT OF NON-NUCLEAR SECURITY OPTIONS IN LIGHT OF RECENT TRENDS

The immediately foregoing discussion affords concrete evidence of major progress on the part of history's two most powerful adversaries in expanding their military cooperation into areas of security thinking and planning that traditionally have not been central to the military profession. For the US military, this progress is being pursued in an effort to complement rather than compete with negotiators in the Department of State and other civilian agencies. Of course, given the ever-present bureaucratic competition for scarce fiscal resources, especially manifest at the present time, the military, in the United States at least, is not likely to be on the forefront pressing for reformist or radical shifts in a security system that it credits with having brokered international peace for the last half-century—and particularly not if such a shift would signal a major decline in the status of the military relative to the maintenance of world peace and order. Nevertheless, the new trends evident in crisis prevention, arms control verification, military information exchange, and military force restructuring clearly suggest a greater receptivity on the part of the professional military to alternative ways of thinking and planning about national and international security—even a global security system that does not rely on nuclear deterrence as its fundamental precept. Indeed, perhaps in combination with "minimum deterrence"—*i.e.*, a *nuclear* strategy in which nations maintain only such nuclear weapons as are necessary to inflict unacceptable damage upon an adversary even after having suffered a nuclear attack—a strong commitment to crisis prevention, arms control verification, military information exchange, and perhaps some limited military force restructuring might be said to constitute a viable security regime worthy of serious pursuit in its own right. But clearly these trends have implications for the viability of those security alternatives that have a realistic chance of implementation *and simultaneously seek escape from nuclearism*. I now return to them for assessment in light of recent developments.

Non-Offensive Defense

The progressive military undertakings initiated in recent years and noted above[67] have made the military establishments of at least the Soviet Union and the United States increasingly receptive to the idea of non-offensive defense (NOD) and possibly even to its place as the principal linchpin of a non-nuclear global security system. Recent successes in crisis prevention, arms verification, and military information exchange have contributed to the military's willingness to consider the feasibility of "limited options," featuring the elements embraced by NOD. They are very much of a piece with the post-Cold War transition in which we find ourselves and therefore contribute to an unprecedented potential for international security enhancement at this historic time.

As indicated, however, military strategists continue to respond to this non-nuclear security option with trepidation—even if with cautious optimism. For example, many in the military remain unconvinced that there exists a clear difference between offensive and defensive weapons and weapons systems. A case in point is the armored tank. Whereas many advocates of NOD identify the tank as a prime illustration of an offensive weapon or weapon system, most military analysts consider the tank, commonly held far behind the front line and then used only to throw back an invader to behind its own lines, as no more offensive when used in this way than a front line stationary bunker equipped with a maingun capable of firing deep into enemy territory. To most military analysts, that is, it is the *intended* use of a weapon or weapon system rather than its *capability* that makes it offensive or defensive in nature. Accordingly, most US (and a great majority of foreign) military analysts reject altogether the idea of labeling entire categories of conventional weapons and weapons systems as offensive to justify their wholesale elimination for the sake of some preferred security regime. Such labeling, the logic of this analysis contends, would cause military technology to concentrate its efforts at the outer edge of the defensive paradigm and thereby increase competition among rival militaries to spend enormous sums in research and development and in doctrinal retraining as each clever breakthrough would come to redefine the manner—in reality, quasi-offensive—in which military forces would conduct themselves. In addition, any unilateral failure to participate in this race would prove more destabilizing to the military balance than "playing the game"; one side's breakthrough unmatched by the other would entice the former to exploit its new capabilities before the latter could close the gap, thus increasing the dangers of sudden and destructive warfare.[68] Moreover, unless accompanied by exceedingly rigorous weapons production verification in all adversary countries, the elimination of a particular weapon or weapon system simply on the debatable grounds that it is offensive would make the covert acquisition of that weapon or weapon system by an adversary highly probable and thus likewise invite instability and the prospect of war. In sum, little or nothing would be gained either financially or militarily from NOD, so its critics contend, and conceivably the potential danger to peace would be greater than under even the present nuclear deterrence system— a point of view apparently shared by some in the Soviet military hierarchy even though they tend to be more subdued in their protests against "weapons-branding."[69]

On the other hand, as noted earlier, there is growing interest in the NOD option as a blueprint theory to justify the pursuit of bilateral quantitative cuts in certain weapons systems that happen to be qualitatively incompatible with NOD.[70] The present-day buzzword, "defensive defense," can be used to instruct the *quantitative* reduction of weapons and weapons systems that exhibit certain *qualitative* characteristics rather than to mandate their elimination altogether. Indeed, judging from the above-noted progressive undertakings of the superpower militaries since 1987 and the probability of reciprocal reductions in conventional weapons and weapons systems between them in the future, there is considerable reason to believe that the military will actively support the *quantitative* disarmament approach to NOD. True, it may be argued that this rhetorical use of NOD amounts to no more than a new name for a less radical pattern of conventional arms reductions. Yet, to the extent that it enables the US, USSR, and other militaries to ascribe a vision to the reduction of particularly expensive and aggressive conventional weapons and weapons systems, it entails a dynamic that can be expected to gain and maintain military support. Also, there is no escaping that these initial ventures into feasible conventional force restructuring might some day soon, when the conditions are right, facilitate more radical strides in the direction of a pure NOD security structure.

Thus, even though the military's commitment to NOD is essentially a commitment to a limited quantitative "draw down" of certain expensive and expansive systems rather than an across-the-board dismantling of offensive forces that wholesale qualitative disarmament or "pure defense" would entail, it represents a

significant departure from the past.   Limited quantitative
conventional force restructuring, complemented by genuine military
cooperation and dialogue relative to crisis prevention, arms control
verification, and military information exchange, today offers the
possibility of a non-nuclear security regime that heretofore could
only have been imagined in the distant future.

Civilian-Based Defense

Civilian-based defense (CBD), which has been little discussed
in this chapter, is given even less attention in professional military
circles. The reason is simple. CBD, as principally a civilian-based
option, allows scarcely any room for the professional military in its
conception or execution. Advocates of CBD assign no role to the
professional military outside of that incumbent in a short-term
"transarmament" phase;[71] and the military, seeing no obvious role for
itself in such a regime, wastes little of its time conjecturing how it
might fit in. The military, in fact, can be counted to stand opposed
to any CBD plans for the most part, given that adoption of such an
alternative to nuclear deterrence would lead to the professional
military's bureaucratic demise.

But participatory issues aside, the military objects to CBD on
two additional grounds. It sees CBD as discounting the possibility
that the use or threat of use of force can have valuable political
leverage for a defending government; and it is concerned that a
determined, force-possessing invader could easily subdue even the
most well organized civilian force.

Regarding the first objection, CBD is a deterrent strategy only,
authorizing a defending government neither a rationale for
warfighting capability nor warfighting wherewithal. The military
establishment, however, is committed to the maintenance of both
deterrence and warfighting capability as essential to a viable national
security strategy.[72]   Furthermore, as many students of international
affairs are quick to note, individual States, particularly the world's
major powers, want to preserve their ability to project military
power, having security interests that are not always amenable to
political or economic leverage in situations beyond their contiguous
borders.[73] Thus, CBD will continue to be seen by the military, in the
foreseeable future at least, as unacceptable or naive or both. None
of the US-USSR military initiatives undertaken in the last few years
belie this conclusion.   CBD still is too ambitious for current
progressive military thinking.

Regarding the second objection, if CBD is to be successful, the defending population probably will suffer in the extreme. History is replete with examples of the effective and ruthless application of force by a cunning invader to break the will and spirit of a determined and proud people. For every Yugoslavia, there is a Quisling Norway, a Vichy France, or a Nazi razing of the Warsaw Ghetto. Added to which, the potential value of CBD is called seriously into question when one critically analyzes the purported successes it has enjoyed according to its proponents. A strong case can be made, for example, that Mahatma Gandhi and his followers would have been totally exterminated had their adversary been Nazi Germany rather than the British Empire.[74] Faced with the threat of an extremely malevolent adversary, the majority of the world still would rather risk biological extinction in war than cultural extinction in a pseudo-peace promised through CBD.[75] Finally, military professionals ask, how is even the best CBD to guard against aggression – particularly aerial or missile attack – that invades not to acquire territory or personal domination but simply to do damage?

These two major objections never are really overcome by the proponents of CBD. Of course, it is at least arguable that an intermediate version of CBD, exhibiting a mixture of military and civilian deterrence components, might win some military support. Utilizing military expertise in, say, the crisis prevention, arms verification, and military information exchange areas is bound to prove more appealing to the military mind than ruling out the military altogether. Even such an intermediate CBD arrangement, however, is unlikely to escape the two principal objections noted above. Doubtless of the view that only a fundamental change in human aggressiveness is likely to guarantee complete freedom from external military threat, even the Swedes and the Swiss retain a warfighting capability. Moreover, it is simply unrealistic to expect that the military, generally conservative in character, is likely to be much interested in any alternative to nuclear deterrence that by definition threatens its own autonomy and viability. At best, CBD is seen by the military as a supplement to, not a substitute for, military deterrence and warfighting capabilities.

An Independent International Peacekeeping Force

Recent US-USSR military contacts have formalized some limited procedures that are compatible with the establishment of international peacekeeping forces, and continued dialogue by way of the joint-military committee to further refine and expand upon these procedures is in the offing.[76] Yet these procedural adjustments are not intended to bring about the establishment of a global armed force along the lines of the Clark-Sohn plan.[77] Rather, they are designed to permit the current problem-plagued peacekeeping operations of the United Nations to remain afloat. Indeed, the most that informed observers believe could come of this process would be a UN peacekeeping force manned concurrently by Soviet and US forces—assuming, that is, that joint US-USSR working groups could establish formal procedures at a reasonable pace over the next decade.[78] Even then, however, large doubts remain. Most nation-state politicians fear the potential *supra*nationalism of such a force and national military officials are genuinely terrified of their loss of autonomy should such an entity come into being. The only precedents that come close to the idea of an independent international peacekeeping force (during the Holy Roman Empire, for example) leave bad images of corruption and abuse in the minds of contemporary strategists. Potentially huge supranational structures evoke the notion that "absolute power corrupts absolutely."

In sum, excepting moderate progress toward more-or-less coincidental involvement in UN peacekeeping ventures, there is little support to be found among the military for independent peacekeeping forces, be they global or regional. The true internationalization of the use of force will come about only after the establishment of international institutions capable of effectively allocating resources and effectively meting out necessary punishment as required.[79] At the present time, with the United States fearful even of adhering to the compulsory jurisdiction of the World Court, the prospects for this alternative to nuclear deterrence look bleak.

CONCLUSION

Though written from a less conservative perspective than most military writings, this chapter is premised on the belief that no realistic alternative to nuclear deterrence can or should evolve without the active engagement of the world's militaries in its definition and execution. As former JCS Chairman Admiral William Crowe put it in June 1989, relative to the "imperative" of the Soviet and US militaries contributing to the reduction of tension and promotion of stability, "[t]here are many technical questions to be sorted out which can only be answered by those with arms expertise. . . . [S]ound judgments by our military leaders will be essential."[80]

Of course, this point of view must be disillusioning to those who think that real security, national or international, is best achieved without the presence or influence of the military in any way. Yet if one is serious about establishing a non-nuclear global security system in the relatively near term, one is well advised not to dismiss the military summarily nor to overlook the genuinely progressive contributions it actually can make toward international security enhancement. Especially at this particular historical time of fundamental change, one must guard against rejecting a glass half full because it is seen as half empty. The historically aware — and prudent — analyst will recall that only a few short years ago the glass was in fact empty altogether!

Taking our cue, then, from the glass half full, which is to say the cooperative undertakings of the Soviet and US militaries in the last few years, not to mention the profound transformations now taking place in Eastern Europe, it is possible to discern a "window of vulnerability" in military conservatism that can be turned to handsome progressive advantage. That is, it is reasonable to conclude that an alternative to nuclear deterrence in the form of restructured military forces designed to foster mutual confidence and crisis stability at militarily verifiable lower conventional force levels is realistically at hand, with the military retaining a decreased warfighting capability, on the one hand, but being authorized an enhanced diplomatic-political role, on the other.

Thus, mindful that historic opportunities await us and that they must be seized lest for some reason they slip irretrievably away, I urgently and strongly recommend a non-nuclear global security regime in the form of a *modified* NOD strategy of national force restructuring, along the lines heretofore indicated, coupled with and featuring the positive interaction of military ideas and personnel as the prime inhibition to conventional warfare. The rough features of this regime would embrace, first, a worldwide web of bilateral and multilateral interchanges among the world's militaries, some of which are being sketched between the Soviet Union and the United States even as I write:

1. joint military working groups targeting crisis limitation procedures and formal, regularized crisis prevention conferences;
2. complete and timely conventional military exercise notification to facilitate observation of all major military rehearsals and practices;

3. complete exchange of military budgeting and force-planning guidelines for coming fiscal years; and,
4. patterned annual visits by the heads of rival military establishments to the full array of the opposite side's military forces, schools, and production facilities.

Of course, while going a long way toward lowering the threshold of fear worldwide and thereby enhancing global security, such a web of military-to-military cooperation would not be enough. It would have to be supplemented by a broad sweep of color and texture featuring a variety of additional cooperative programs and procedures, among them probably the following:

5. regularized joint military strategy and doctrinal conferences that seek to align, in a stable manner, both the military force posture and the military doctrines of the involved parties;
6. abundant and regular exchanges of military officer *students* at all levels of military schooling among the involved countries;
7. equally abundant and regular exchanges of military officer *instructors* at all levels of military schooling conducted among the involved countries;
8. institutionalized joint military working groups chartered to assess the cost and potential dangers of emerging technologies with conventional weapons spillover, and then to make simultaneous recommendations to their respective national leadership as to those technologies and/or weapons that would be better negotiated away before a costly and destabilizing spiral competition in that area ensues;
9. the mandatory exchange of a contingent of military personnel of all ranks and services for all major adversarial training exercises; and,
10. the continued press of each military, by way of the joint working-group framework established, to train for and conduct the bulk of the conventional weapons control verification required for the maintenance of a stable and peaceful conventional security regime.

Anchored as these ten proposals are in increased cooperative contact and reciprocal monitoring among the world's militaries, they hold out the potential for significantly enhanced national and international non-nuclear security and therefore should be pursued without delay. Even more progressive alternatives to nuclear deterrence could be pursued thereafter, capitalizing upon growing military competencies in crisis prevention, conventional arms control verification, military information exchange, and conventional force restructuring.

This recommended approach, surely as worthy of consideration as the other non-nuclear security options outlined in this chapter, seeks to modify rather than eliminate the role of the military in the attainment of post-nuclear global security. It does so for the simple reason that, despite the apparent end of the Cold War and other positive developments, it is by no means clear that the world as a whole is prepared to give up war itself as an instrument of national policy, permitting the abolition of sovereign military forces in favor of a broadly effective international legal order or some supranational system of global governance. Deferring to the military's interest in a full albeit redefined participatory role, it minimizes not only the risks of catastrophic conventional war but also the risk of nuclear war, ever a possibility where there is a will.

The real danger to world peace and security, I submit, lies in the possibility of failing to capitalize on present progressive tendencies within the military and other historic opportunities either by insisting upon outmoded absolutist notions of national power and prestige or by insisting that any change must eliminate all traces of militarism. However, taking appropriate advantage of the present historic moment, along the lines recommended here, can enable genuine progress toward a new non-nuclear security regime between the US and the USSR in the near term and, by the dawn of the Twenty-first Century, a new non-nuclear security regime and greatly restrained militarism for the rest of the world as well.

NOTES

The opinions expressed by the author are entirely his own, except as noted. They are in no way intended to express or represent the policy of the Department of Defense or the United States Army relative to the issues discussed.

1. Steven Meyer, "The Sources and Prospects of Gorbachev's New Political Thinking on Security," *International Security* 13, 2 (Fall 1988): 139-40.
2. *See* Bruce Parrot, "Soviet National Security Under Gorbachev," *The Problems of Communism* (November-December 1988): 1-36.
3. Mikhail Gorbachev, *Perestroika: New Thinking for Our Country and the World* (New York: Harper and Row, 1988): 184.

4. Treaty Between the United States of America and the Union of Soviet Socialist Republics on the Elimination of their Intermediate-Range and Shorter-Range Missiles, December 8, 1987, Dep't State Pub. 9555 (December 1987), *reprinted in* Burns H. Weston, Richard A. Falk, & Anthony D'Amato (eds.), *Basic Documents in International Law and World Order* (St. Paul, MN: West Publishing, Second Edition, 1990): 279-90.

5. In a development continuing the momentum of the two superpowers toward the reduction of their nuclear weapons stockpiles, Soviet Foreign Minister Eduard Shevardnadze told the opening session of the UN General Assembly that he anticipated that "the last turn on the road toward a treaty reducing strategic offensive arms [should be passed before Summer 1990]." He also indicated that the Soviets believed that the next step in the denuclearization process between Moscow and Washington should feature (1) the phasing-out of production of all fissionable materials used in nuclear explosives, (2) new restrictions on nuclear tests, and (3) steps toward halting the spread of missile technology necessary for credible nuclear delivery systems. *See* Paul Lewis, "Soviets, Welcoming Bush's Plan on Chemical Arms, Go Further," *The New York Times* (September 27, 1989): A-12.

6. There are numerous excellent works on the essential components of national and international security, many of them taking differing positions relative to what those components are. The approach taken here is purposefully a military-oriented conception. Its principal tenets and precepts are outlined in a more detailed manner in Samuel P. Huntington, "Conventional Deterrence and Conventional Retaliation in Europe," *reprinted in* Steven E. Miller (ed.), *Conventional Forces and American Defense Policy* (Princeton, NJ: Princeton University Press, 1986): 251-75.

7. *See especially* Fred S. Hoffman, "The SDI in U.S. Nuclear Strategy," and James R. Schlesinger, "Rhetoric and Realities in the Star Wars Debate," in Steven E. Miller and Stephen Van Evera (eds.), *The Star Wars Controversy* (Princeton, NJ: Princeton University Press, 1986): 3-24.

8. *See* Samuel P. Huntington, *The Soldier and the State* (Cambridge, MA: Harvard University Press, 1957): 3-67, a seminal treatise on civil-military relations.

9. *See* John L. Gaddis, *Strategies of Containment* (Oxford: Oxford University Press, 1982): especially 3-163; David Allen Rosenberg, "The Origins of Overkill," in Steven E. Miller (ed.), *Strategy and Nuclear Deterrence* (Princeton, NJ: Princeton University Press, 1984): 113-82.

10. *See* Gorbachev, *Perestroika*: 181.

11. The complete text of the speech may be found in *Presidential Documents* 19, 12 (Monday, March 23, 1983).

12. *See* the excerpts from the Hoffman and Fletcher reports in Miller and Van Evera, *Star Wars Controversy*: 273-89.

13. For pertinent discussion, see the chapter by Warren F. Davis in this volume.

14. For a complete review of the technical problems with the SDI proposal, see especially Charles L. Glaser ("Why Even Good Defenses May

Be Bad" and "Do We Want the Missile Defenses We Can Build?") and Drell, Farley and Holloway ("Preserving the ABM Treaty: A Critique of the Reagan Strategic Defense Initiative") in Miller and Van Evera, *Star Wars Controversy*: 25-129.

15. Treaty Between the United States of America and the Union of Soviet Socialist Republics on the Limitation of Anti-Ballistic Missile Systems, May 26, 1972, 23 U.S.T. 3435, T.I.A.S. No. 7503, *reprinted in* Weston, Falk, and D'Amato, *Basic Documents*: 213-15.

16. President Bush has so far abstained from any direct reference to SDI as an impenetrable anti-nuclear shield, and he and his advisors appear also to have downplayed its viability as a "bargaining chip" for future nuclear negotiations. *See* Michael R. Gordon, "Star Wars Fading as a Major Element of U.S. Strategy," *The New York Times* (September 28, 1989): A-I, 11.

17. The term "defensive defense" is perhaps most in vogue as of this writing because of its use by President Gorbachev in his writings and official pronouncements. However, the term "non-offensive defense," or NOD, is preferred here because of its clear juxtaposition of the military strategies of offense and defense. In the literature, it is used interchangeably with the term "non-provocative defense" and sometimes also with the term "qualitative disarmament."

18. *See, e.g.*, Anders Boserup, "A Way to Undermine Hostility," *Bulletin of the Atomic Scientists* 44, 7 (September 1988): 16; Horst Ahfeldt, "New Policies, Old Fears," *ibid.*: 24. These two essays are part of a symposium on non-offensive defense, embracing articles on diverse aspects of the topic by nine other analysts. The symposium may be found at pages 12-54 of the *Bulletin*.

19. *But see* notes 17 and 18.

20. This discussion of the key tenets of non-offensive, or defensive defense, is set out in greater summary detail in a recent useful work: Harry B. Hollins, Averill L. Powers, and Mark Sommer, *The Conquest of War: Alternative Strategies for Global Security* (Boulder, CO: Westview, 1989): 78-84. *See also* Dietrich Fischer and Alan Bloomgarden, "Non-Offensive Defense," *Peace Review* 1, 2 (Spring 1989): 7-11.

21. As delineated in "A Menu of European Defense Plans," in *Bulletin of the Atomic Scientists* 44, 7 (September 1988): 23.

22. For sympathetic comment, see Andrzej Karkoszka, "Merits of the Jaruzelski Plan," *Bulletin of the Atomic Scientists* 44, 7 (September 1988): 32.

23. *Ibid.*

24. The complete text of the Gorbachev address may be found in 43 U.N. GAOR (72nd mtg.) at 2, U.N. Doc. A/43/PV.72 (1988).

25. Gene Sharp, *Making Europe Unconquerable: The Potential of Civilian-Based Deterrence and Defense* (London: Taylor & Francis, 1985): 114. *See also* Anders Boserup and Andrew Mack, *War Without Weapons* (New York: Schocken Books, 1975); Richard Fogg, *Non-military Defense Against Nuclear Threateners and Attackers* (Stevenson, MD: Center for the Study of Conflict,

1983); Adam Roberts, *Civilian Resistance as a National Defense* (Harrisburg, PA: Stackpole, 1968)

26. Sharp, *Making Europe Unconquerable*: 114-20.

27. *See* Alternative Defence Commission, *Defence Without the Bomb* (London: Taylor & Francis, 1984); *Ibid.*, *The Politics of Alternative Defence: A Role for a Non-Nuclear Britain* (London: Paladin, 1987).

28. *See, e.g.*, Gene Sharp, *National Security Through Civilian-Based Defense* (Omaha, NE: Civilian-Based Defense Association [Association for Transarmament Studies], 1985).

29. Hollins, Powers, and Sommer, *Conquest of War*: 38.

30. For the latest version of the plan, complete with a detailed introductory summary indicating its underlying principles and main features, see Grenville Clark and Louis B. Sohn, *World Peace Through World Law* (Cambridge, MA: Harvard University Press, Third Edition Enlarged, 1966). For an abbreviated helpful summary, see Hollins, Powers, and Sommer, *Conquest of War*: ch. 4.

31. *See* Clark and Sohn, *World Peace Through World Law*: Annex, pp. 314-34.

32. For other analyses recognizing the same need, see Robert C. Johansen and Saul H. Mendlovitz, "The Role of Law in the Establishment of a New International Order: A Proposal for a Transnational Police Force," *Alternatives: A Journal of World Policy* 6, 2 (July 1980): 307; and Walter Mills, "The Role of Police in Response to Violations," in Richard A. Falk and Richard J. Barnet (eds.), *Security in Disarmament* (Princeton, NJ: Princeton University Press, 1965): 305-11.

33. For a theoretical discussion of the wisdom of this assertion, see Kenneth Waltz, *The Theory of International Politics* (New York: Random House, 1979): especially chs. 6-9.

34. *See* note 4.

35. *See, e.g.*, Kurt M. Campbell, "The Soldier's Summit," *Foreign Policy* 75 (Summer 1989): 76.

36. *Ibid.*, 76-7; and Francis X. Clines, "US-Soviet Accord Cuts Risk of War," *The New York Times* (June 13, 1989): A-12.

37. At the time of the signing of this protocol, JCS Chairman William Crowe noted that he was especially pleased that military professionals negotiated the pact directly. He expressed the desire to make such protocols a continuing pattern between the two military establishments. For details, see Clines, "US-Soviet Accord."

38. Campbell, "Soldier's Summit": 78.

39. While this is a personally conceived listing of aggregate areas of advancing military interests in non-traditional roles, I am indebted to the organizational framework of the last three pages of Campbell's "Soldiers Summit."

40. As in such inherently hostile environments as the Persian/Arabian Gulf or the waters of the Greenland-Iceland-United Kingdom (GIUK) Gap.

41. Gorbachev, *Perestroika*: 181.

42. Former JCS Chief Admiral William Crowe and Chief of the Soviet General Staff Mikhail Moiseyev signed this agreement on June 12, 1989. To take effect January 1, 1990, the agreement calls for the establishment of a commission of military experts from both countries to study the risks that grow from standard military readiness practices and from such areas as laser-weapons testing, radio jamming, accidental airspace incursions, and the creation of special control areas (such as the Persian/Arabian Gulf). *See* Clines, "US-Soviet Accord": A-12.

43. *See* Alexander L. George, "Crisis Prevention Reexamined," in Alexander L. George (ed.), *Managing the US-Soviet Rivalry* (Boulder, CO: Westview, 1983): 335-64.

44. *See Department of Defense Annual Report to Congress, FY 1990* (Washington, DC: U.S. Government Printing Office, January 9, 1989): 46, where the three major issues subjected to the June 1989 protocol were listed as including: (1) the creation and maintenance of a Joint Military Working Group (JMWG) to avoid dangerous military-to-military incidents; (2) the institutionalization of military exchanges to foster mutual understanding and a transfer of the knowledge being built up on the conduct of defensive defense (*see* discussion *infra*); and (3) the establishment of increased military-to-military contacts in all feasible areas.

45. Campbell, "Soldier's Summit": 89.

46. For a discussion of the role that limiting arms transfers can have in confidence-building between the two militaries, see Barry Blechman, Jane Nolan, and Alan Platt, "Negotiated Limitations on Arms Transfers: First Steps Toward Crisis Prevention?" in George, *Managing the US-Soviet Rivalry*: 255-84.

47. For greater discussion, see Editorial, "Controlling Risks, Not Just Arms," *The New York Times* (June 14, 1989): A-28, wherein the *Times* editors describe how procedures for weapons accountability and security exchanged between the two militaries can help one side to critically review and adjust internal procedures based upon the ideas of the other.

48. *See* note 4.

49. Campbell, "Soldier's Summit": 85.

50. As detailed in *ibid.*, 79.

51. For a concise history of the Stockholm Conference and its results, see John Borawski *et al.*, "The Stockholm Agreement of 1986," *Orbis* 30, 4 (Winter 1987): 643-63.

52. *Ibid. See also The Iceland Proposals: Implications for National Security* (West Point, NY: A USMA White Paper, 1986).

53. Campbell, "Soldier's Summit": 90.

54. *See* Robert Pear, "U.S. and Moscow Settle the Key Issues on Chemical Arms," *The New York Times* (July 18, 1989): A-I, A-6.

55. *See* this chapter: Recent Progressive Military Undertakings.

56. *See* Hiroyuki Agawa, *The Reluctant Admiral: Yamamoto and the Imperial Japanese Navy* (New York: Harper and Row, John Bester Translation, 1979): 15-72.

57. My conclusion is derived from an unclassified State Department cable, dated July 12, 1989, incorporating the plenary statement of the chairman of the US Delegation to the CSCE working group on Conventional Security Balancing Measures (CSBM) wherein the chairman (Maresca) stated: "I believe we can now say that there is a broad agreement in principle in this forum on a comprehensive annual information exchange. In addition, there appears to be an agreement in principle to the concept of an information exchange on planned major conventional weapon deployment programs. [This] agreement in principle is clearly a major step forward for confidence-building in Europe, [and] this is especially true in comparison to the situation only three years ago, in Stockholm, when an agreement on the exchange of information was not possible."

58. I especially refer to remarks attributed to the Allied Commander of all United States and Korean forces in South Korea, General Louis C. Menetrey, in the August 14, 1989, issue of *The Army Times*. According to this news report, this former staunch advocate of a continued US presence in the Republic of Korea (up to at least the 48,000 troops reported there at the end of 1988) moderated his position by indicating that he would find a gradual "draw down" of US forces from Korea in the 1990s to be reasonable and prudent. General Menetrey indicated that a US withdrawal was now likely due to President Bush's May 1989 proposal to withdraw 30,000 troops from Europe, the continuing improvement of the South Korean economy and military, and the moderating influence of reforms in the USSR and China on the anticipated behavior of the North Koreans.

59. For more information on the details of the Bush offer at the meeting of the North Atlantic Council in Brussels on May 30, 1989, see *The Declaration of the Heads of State and Government Participating in the Meeting of the North Atlantic Council in Brussels* (May 29-30, 1989), published by the NATO Press Service (1110 Brussels, Belgium): 4-5. Also note that an even smaller US force total of 225,000 troops in NATO was tabled in the CFE talks during February 1990.

60. See Gorbachev, *Perestroika*: chs. 2 and 7; and Steve Lohr, "Soviet Economic Bureaucrats Are Not About to Fade Away," *The New York Times* (August 15, 1989): D-5, wherein the author noted that President Gorbachev announced in January 1989 that he planned a military spending cut of 14 percent over the ensuing two years, with the economic resources to be shifted to increased production of consumer goods in the Soviet economy.

61. See this chapter: Non-Offensive Defense.

62. Comments by General (Ret.) Goodpaster spoken in the presence of the author in the course of an informal set of remarks made as part of a panel discussion on emerging security issues at the Woodrow Wilson School at Princeton University on April 8, 1989.

63. The announced focus of the Atlantic Council was made in the final remarks of its June 1989 convention as seen on C-SPAN.

64. Henry Owen and Edward C. Meyer, "Central European Security," *Foreign Affairs* 68, 3 (Summer 1989): 37-38.

65. The word "deep" in this context is, of course, a value-ladened term. Generally, however, it is taken to mean that military forces retain the ability to execute 10-50 kilometer-deep "counterattacks" but not offensive interdictions into adversary territory much beyond a total of 15-20 kilometers. These deeper ground incursions are referred to as "counteroffensives."

66. In addition to the writings of Boserup and Ahfeldt in the *Bulletin of the Atomic Scientists*, see Gorbachev, *Perestroika*: ch. 7; Robert C. Johansen, "Toward an Alternative Security System," in Burns H. Weston (ed.), *Toward Nuclear Disarmament and Global Security: A Search for Alternatives* (Boulder, CO: Westview, 1984): 569; and Hollins, Powers and Sommer, *Conquest of War*.

67. *See* this chapter: Recent Progressive Military Undertakings.

68. *See* Stephen J. Flanagan, "Nonoffensive Defense Is Overrated," *Bulletin of the Atomic Scientists* 44, 7 (September 1988): 48-49.

69. Critics of NOD note that the Soviet military is having a large problem in creating a non-provocative tactical entity in its force structure. Congressional and military delegation visits to defensive military exercises held by the USSR and its Warsaw Pact allies in Summer 1989 have bred more skepticism than enthusiasm. Observers of these exercises note that (1) Soviet pioneering efforts at "purely defensive" operations look strikingly similar to offensive operations but for smaller-size formations; and (2) the promised removal of Soviet tanks – *the* archetypical offensive weapons system according to Soviet theorists – from these defensive tactical testbeds has not been as complete or generous as forecast. *See* Michael Dobbs, "Joint Chiefs View Soviet Weaponry," *Washington Post* (June 19, 1989): A-I; and Michael R. Gordon, "Congress Inspects Soviet Pullback," *The New York Times* (August 9, 1989): A-7. For a discussion of the Soviet military's concerns and machinations over their "forced" adjustment toward the doctrine of non-offensive defense, see Meyer, *Gorbachev's New Political Thinking*: 161.

70. *See* this chapter: Changes in Force Composition and Structure.

71. *See* Sharp, *Making Europe Unconquerable*: 114-20.

72. *Deterrence* is best described as the prevention of aggressive war through the maintenance of a known and credible military capability. *Warfighting capability* is best defined as the demonstrated capacity of a State's military forces to preserve the territorial integrity and political sovereignty of the nation-state against hostile external forces or, alternatively, to threaten the military establishment, territorial sovereignty, or political viability of hostile or potentially hostile States in a manner calculated to prevent attack. In the military mind, the objective of conventional (and, to much lesser extent, nuclear) deterrence cannot be securely maintained without parallel maintenance of a warfighting capability and *vice versa*; they are seen as mutually reinforcing.

73. This view in US military thinking dates to the turn of the century and was cemented by World War II. The professional military believes ardently that the need to maintain a "power projection posture" to protect US political interests in overseas locations is a non-negotiable mandate of modern security systems, a requirement that cannot be met by even the militia system much

less strict CBD. This position would seem irreversible until the world community establishes international institutions that can better protect US overseas interests than the maintenance of a strong, sovereign military "power projection" capability. For a detailed discussion of this US military view in the Twentieth Century, *see* Jim Dan Hill, *The Minute Man in Peace and War: A History of the National Guard* (Harrisburg, PA: Stackpole, 1964): chs. 7-9.

74. *See* Hollins, Power, and Sommer, *Conquest of War:* 95-96.

75. *See* Benjamin Cohen, *The Quest of Imperialism: The Political Economy of Dominance and Dependence* (New York: Basic Books, 1973): 237.

76. *See* Campbell, "Soldier's Summit": 86-87.

77. *See* this chapter: An Independent International Peacekeeping Force.

78. *See* Campbell, "Soldier's Summit": 86-87.

79. *See* Cohen, *Quest of Imperialism*: 251-57.

80. From Admiral Crowe's path-breaking speech to Soviet military academy students, reproduced in the *Wireless File* (June 15, 1989), a publication of the US Department of State.

2

Technology and Alternative Security: A Cherished Myth Expires

Warren F. Davis

Heaven forbid that I should be led into giving countenance to superstition by a passion for impartiality, and so come to share the fate of Eusapia's patrons!
— H.G. Wells, *The Plattner Story*

Almost without exception, complex and difficult problems demand complex and difficult solutions. Yet, ever present is the powerful temptation to seek the easy way out. This tendency is no less true in the realm of national and international security than it is in any other. Indeed, it may at least partially explain the lingering support that continues for the US government's Strategic Defense Initiative (SDI), or "Star Wars" program, proposed by President Reagan in 1983.[1] A very limited—and frankly myopic— view of the relationship between technology and security equates the mere quantity or sophistication of the former with the (presumed) enhancement of the latter. How tempting the notion that we will somehow be "home free" if we just can get ourselves around the next technological corner! The fact is, however, that no technology or cluster of technologies can alone ensure genuine security, national or international, in this nuclear age.

Eleven days after the bombing of Hiroshima, Secretary of War Henry L. Stimson, together with nuclear physicists J. Robert Oppenheimer, Arthur H.Compton, Enrico Fermi, and Ernest O. Lawrence, made the following plea:

> We have been unable to devise or propose effective military countermeasures for atomic weapons. Although we realize that future work may reveal possibilities at present obscure to us, it is our firm opinion that no military countermeasures will be found which will be adequately effective in preventing the delivery of atomic weapons. . . . We believe that the safety of this nation—as opposed to its ability to inflict damage on an enemy power—cannot be wholly or even primarily in its technical prowess. It can only be based on making future wars impossible. . . .[2]

Forty-five years have not diminished the force of this remarkable plea. Indeed, the futility of a guaranteed technological defense against nuclear weapons has been proven repeatedly through almost a half-century of military science. Despite the relentless pursuit of virtually every military technology feasible since World War II, especially on the part of the superpowers, the world—the superpowers included—is now far less secure than it was when it entered seriously into the arms race in the early 1940s. This is not to say that technology can in no way contribute to the security of a post-nuclear world that, though having abandoned nuclear weapons, is nonetheless capable of "reinventing" them virtually at will. Rather, it is to recognize that, in such a world, new security technologies are at most only a part of the solution, and probably not a very large part at that. It is to recognize that the link between technology and genuine security—military, environmental, economic, and so forth—has more to do with the process by which we make technology decisions relating to security than it does with the invention of new and fancier technological hardware.

FUNDAMENTAL LIMITS INHERENT IN THE APPLICATION OF TECHNOLOGY TO SECURITY

An early scene in Stanley Kubrick's film *2001: A Space Odyssey* depicts the moment at which an ancestral hominid first conceived of, and used, a rudimentary weapon: a femoral bone with which a member of the same species is bludgeoned to death over competition for food and water. The bone is subsequently thrown with exultation high into the air and thereafter left to tumble to Earth, whereupon it melds into the image of an advanced spacecraft en route to the moon. In this brief moment of cinematic genius, we see the intertwining of weaponry and technology through the millennia to the present, the failure of restraint in the exercise of power expanded artificially beyond the limits set by evolution, the intoxication evinced by its savage application, and the oblivious disregard for future consequences. In this apocryphal scene, we witness the uniquely human conflict between instinct and intellect that has brought humanity to the brink of self-extinction.

Kubrick's triumphant hominids doubtless assumed that, through their new-found weapons technology, they would enjoy a decided advantage indefinitely. But such assumptions never have been borne out in practice and probably never will be. The opposition is human too and, like it or not, has generally the same potential to discover and apply new weapons technologies—an annoying detail that typically is dismissed via a repertoire of self-deceptions that define the opposition as intellectually, culturally, and socially inferior, as sub-human. Moreover, breakthroughs that come about more by chance than in response to intellectual effort have the same probability of benefiting one adversary as another. But again we are inclined to deceive ourselves. God favors us, we say, because our cause is just. He can be relied upon, we insist, to discriminate in our favor in the meting out of chance discoveries.

Such rationalizations, though an indispensable part of the arms race, seem silly when written down. And indeed they are! For were they true, and were it so that lasting security could be achieved by technological means, some generation since Kubrick's hominids would have long ago gotten a permanent leg up on the opposition and held it there. Plainly, lasting security never has been and never will be achieved through technology alone. The other side will learn soon enough how to use a discarded bone as a club and can be expected to improve upon the idea as well. Technological challenges tend to have technological solutions and technological solutions tend to be found by technological beings. Only by debasing the opposition and by invoking Divine Providence can the illusion of security be maintained. It is a tribute to the power of ideology (or myth) over reason that otherwise supremely rational beings can cling so tenaciously to gut instincts that fly in the face of the historical evidence. Nothing in the history of human affairs supports the assumption that the balance of nuclear terror or any other military standoff can be maintained indefinitely, the conduct of the arms race since 1945 notwithstanding.

The point especially to be emphasized, however, is that a world free of nuclear weapons, in contradistinction to a completely disarmed one, is no more capable of ensuring lasting security by technological means than is a nuclearized world. The "vertical" as well as the "horizontal" proliferation of the arms race since 1945, in terms of *both* conventional and nuclear weapons and weapons systems, has brought to the last half of the Twentieth Century whole classes of problems never encountered by our forebears and for which, perforce, we have little to go on by way of historical guidance.

For one thing, through the relentless application of ever more sophisticated technology to weaponry, the time available to react to perceived threats has decreased dramatically. Intercontinental ballistic missiles, which in a post-nuclear world could carry conventional weapons of mass destruction, are capable of reaching the opposition's territory within half an hour and cannot be recalled after launch nor destroyed in flight by remote control. By design, they respond to no signals, admitting no possibility of the opposition sabotaging an attack. Submarine-launched ballistic missiles that can be brought by sea close to the enemy's homeland decrease the available reaction time yet further to the order of five to ten minutes. And the proposed "Star Wars" technology, which could be used against non-nuclear as well as nuclear missile payloads, could reduce the effective reaction time to a matter of seconds — the launch vehicle's "boost burn" time.[3] The time available under SDI might be essentially zero if the concomitant ASAT (anti-satellite) capabilities of SDI are utilized in initiating an attack.[4]

Additionally, the rate at which human beings can assimilate, analyze, and act upon information has not undergone an improvement corresponding to the improvement in the technology that has been applied to weaponry. Inevitably, more and more contingencies have had to be anticipated, and the reactions thought appropriate have had to be relegated to automatic procedures and programmed systems. The more the technological "thumbscrews" are turned, the greater must be the reliance on automated systems and the lesser will be the role of the human being. Thus, the US Department of Defense seeks now to develop computers with "artificial intelligence," with the capacity to think like humans, one more illustration of the extremes to which evidently we are willing to go rather than face the underlying problem.[5]

Another modern-day problem inextricably linked to advanced technology in service to security by force is the "mutual garrison" phenomenon. Just as the progression in weapons technology has brought an unwelcome decrease in available reaction time, so also has it come to blur the distinction between "their turf" and "our turf."

Strategic bombers built to deliver weapons of mass destruction have so far been confined essentially to mutually exclusive air spaces. The situation is quite different relative to missile-outfitted submarines that navigate in a shared operational area of over 40 million square miles, the better part of the world's accessible ocean surface.[6]

The problem is likely to be exacerbated in the high technology frontier of space if SDI ever is deployed. Here the laws of nature endow the concept of space-based defense with some rather unwelcome and, as far as one can see, intractable problems and characteristics. One of the most serious is rooted in the very nature of earth orbit itself. Everything is in a state of continuous motion on a global scale and everything comes close to everything else sooner or later. Just as serious is the fact that space is, by its nature, empty; there is nothing to hide behind or within and nothing to use as camouflage. Space-based assets are therefore inherently vulnerable in the extreme. Virtually all proposed space-based SDI components, across the entire spectrum of technologies, from kinetic-kill weapons to exotic laser systems, and whether for the near or long term, suffer from the profound defect that they are vulnerable to comparatively crude—hence cost-effective—countermeasures such as orbiting "space mines." The current vogue among some SDI proponents is to suggest, in response to this problem, the maintenance of "keep-out zones" around critical orbiting assets,[7] which is a little like trying to prevent the rain from falling through an eight-foot hole in your roof by declaring the inside of your house a keep-out zone. Nature does not oblige. True, some cling to the belief that technology can always find a way out of any dilemma. But, characteristically, each proposed Band-Aid leads directly to another insuperable problem.

The difficulty of maintaining keep-out zones is compounded by the presence of tens of thousands of objects of varying sizes in orbit around the earth. None have the fuel and control systems that would be required to keep them out of specified zones, and the task of monitoring them is staggering. At best, all objects launched after a certain date could be required, assuming the assent of all States with access to space, to maintain stations outside the forbidden zones. But this leads immediately to another problem: Sooner or later all such station-keeping systems must fail, leaving the associated objects free to drift into the keep-out zones. Though some have recognized that major breakthroughs may be required if the SDI is ever to be feasible, it is hardly reasonable that this should be understood to include the development of space systems with infinite lifetimes. Recognition of this dubious requirement has spawned the proposal that, instead of all other objects in space avoiding the vicinity of the critical assets, the critical assets themselves will be required to avoid all other objects. But this proposal leads directly to yet another absurdity. It requires that the objects to be protected be capable, on their own, of moving away from other objects that approach too closely. Since the SDI components in question would be among the most massive in space, there is no plausible way that the approach of a relatively fast-moving space mine or torpedo could be avoided by such a "step aside" maneuver. Newton's law of inertia cannot be revoked. The idea of automatically destroying anything that drifts into a keep-out zone is equally absurd, even if it were feasible.[8]   Imagine the sky being swept clean of thousands of expensive commercial and research satellites, to say nothing of hapless astronauts adrift due to a malfunction, in the context of an automated system that can allow no external override for fear that the opposition might discover how to turn it to its own advantage. And so it goes.

The common thread that binds this second class of problems is a finite environment permeated by ever more sophisticated weapons. Inevitably the protagonists find themselves defending against each other from within the same garrison, the mutual garrison phenomenon. It is a quirk of locally Euclidian geometry, which will not yield for "man nor beast," that brings with it a plethora of insuperable problems of the most fundamental kind.

Finally, a phenomenon somewhat less general than those considered so far but no less inescapable – and with comparable potential to frustrate the goal of security through technological means – is derived from a principle I call "the affinity of complexity and obsolescence," applicable primarily to very large integrated technological systems.   A prominent characteristic of successive arms race proposals is that they tend toward ever more expansive distributed systems; greater and greater numbers and varieties of components and subsystems are required to realize the overall system mission.   This trend is a natural and unavoidable consequence of the parallel trend, already noted, toward progressively more automated globally comprehensive systems of which SDI is so far the ultimate example.   As system complexity increases, the problem of systems integration becomes correspondingly more difficult.[9] With the advent of SDI we have reached the point where the difficulty of the integration problem begins to dominate all other technological problems, acute though they may be.[10] The conundrum that attaches to the "affinity of complexity and obsolescence" principle transcends even the system integration problem and lies in waiting should defense managers ever manage successfully to navigate the integration waters.

As system complexity increases, it becomes necessary earlier and earlier in the development phase to commit to technologies that will be used in the completed system. This is unavoidable because the ramifications associated with changes of technology become more and more difficult to isolate and deal with as overall complexity increases. And, of course, added difficulty translates into added time, so that the penalty for the accommodation of newer technologies in midstream is, generally, postponement of the project completion date and escalating costs by increments that increase with complexity. There emerges a tradeoff between the ability to introduce newer technological developments in midstream and the ability to deliver the final operational system. The less complex the system, the more readily changes can be accommodated and the more closely the final product can reflect the technological state of the art. Conversely, the more complex the system, the less readily can changes be accommodated, with the result that the technologies embodied in the final product tend more and more toward obsolescence.[11] Hence: the affinity of complexity and obsolescence.

All of which bodes ill for anyone still taking SDI or comparable schemes seriously. Considering its unprecedented complexity, it may well have to be deployed with technology twenty-five or more years out of date. And, after initial deployment, the extraordinary difficulty of retrofitting modifications, both in terms of cost and of assuring the maintenance of reliable systems integration, will bias SDI toward even greater obsolescence. Extremely complex distributed systems such as SDI are born with an exposed Achilles tendon, whereas effective countermeasures, being required only to foil the opponent's systems by any means (however lacking in sophistication), enjoy the advantages of reliability and cost-effectiveness that attend considerably reduced complexity. Ironically, the hope for security through technological means becomes less and less realistic as systems become more and more comprehensive and complex.

Thus we see, in the preceding paragraphs, some of the intrinsic factors that limit humankind's ability to achieve security indefinitely through technological means. It should be noted that none of these factors derive from the particulars of the systems or techniques involved; they are intrinsic, arising independently of specific concepts, and have become evident only recently due to the level of technological complexity and sophistication that, at the close of the Twentieth Century, we finally have been able to achieve. Though these limits are an anathema to persons bound by traditional ideas of what ought to be possible, their fundamental character and foreboding implications can be ignored only at our peril.

SOME HELPFUL TECHNOLOGICAL OPTIONS

Despite the fundamental limitations that make impossible the achievement and maintenance of global security exclusively through technological means, technology can make at least a few, possibly even some indispensable, contributions to the achievement and maintenance of an essentially nuclear-weapons-free world. Just as technology has contributed to the arms race—the *raison d'être* of which resides, at bottom, with technology—so also might technology contribute to the *peace race*. In particular, technology can be helpful insofar as it assists the world community and its constituent polities to become more *anticipatory* and *preventive* rather than merely *reactive* in the execution of security strategy. The menace embodied in nuclearism is of such character that technological and other security measures that fail to anticipate and prevent are not especially valuable. Thus, technologies that focus on the verification and enforcement of nuclear arms reductions and on the monitoring of relative differences in conventional forces would appear promising. Of course, other as yet unknown technologies surely will emerge in the years to come to assist the promotion and protection of a post-nuclear world.

While we consider suggestions for technologies that could assist a post-nuclear global security system, however, it must be borne in mind that the very idea of such technologies makes sense only in the context of an overriding geopolitical regime that is dedicated to the permanent end of arms racing. There simply is no hope of lasting and meaningful security through technological means alone. Indeed, as discussed above, it is advanced technology itself that has driven us to the point that any margin of safety temporarily gained is inevitably eroded and then followed by even greater danger than existed at the outset. This is of course precisely the position in which the world finds itself after forty-five years of attempting security through ever more advanced technological proposals and "solutions."

Also, it is important to proceed cautiously in the definition and application of alternative security technologies, especially in the wake of the recent sweeping reforms in the Soviet Union and Eastern Europe and the parallel emergence of strange national and international security bedfellows. Arms controllers and disarmers, formerly critical of defense contractors, now offer helpful defense suggestions; and defense contractors, accustomed to looking upon arms controllers and disarmers with jaundiced eyes, now listen to these suggestions both cordially and attentively. But unless an irreversible shift really does take place, parties with traditionally conflicting goals can become too easily the unwitting servants of the other. In the absence of a fundamental change in international relations, defense contractors could be entering merely into a kind of holding pattern, awaiting the return of a cold or even hot war – in the competition for scarce resources or the battle against drugs, for example – and in the process gaining the acquiescence, even the active cooperation, of their former critics.

Now, with these preliminary caveats in mind, let us turn to those anticipatory and preventive technologies that seem the most promising for a post-nuclear world, bearing in mind that among my present purposes, as shall be seen, is an urgent desire to provoke engineers, physicists, and other technical scientists to reconsider as much as possible their relationship, if any, to the military industrial complex.

Surveillance Satellites

A set of technologies especially well suited to anticipating and preventing resort to force, even small-scale uses of force, is that of orbiting satellites and remote sensors. Both in the East and West, surveillance satellites currently carry a wide array of sensors – photographic cameras, multispectral scanners, infrared sensors, microwave radars, and electronic listening devices – that are capable of obtaining many different types of information from the earth's surface; and while their orbital paths, their operating wavelengths, and the intervening atmosphere impose certain fundamental limits on their effectiveness, they are becoming both increasingly sensitive and increasingly precise.

As part of a complex of safeguards designed to ensure a world free of nuclear weapons, surveillance satellites, available to national and international security managers, could perform a number of critical functions. Most obviously, they could help to guarantee compliance with arms control and arms reduction agreements, exposing violations when they occur, which then could be reported and discussed. They even could be used to spot the deployment of chemical weapons and factories capable of making them.[12] Similarly, they could be used to monitor troop movements and conventional force deployments so that potential conflicts might be addressed at an earlier stage with the goal of preventing an escalation of crisis.[13] Finally, they could be used to facilitate timely and convenient intergovernmental communication.

As early as 1978, the French government proposed to the United Nations General Assembly the establishment of an international satellite monitoring agency that would oversee most if not all of the potential functions of satellite surveillance mentioned above.[14] Such an agency would include an international team of experts that could analyze, interpret, and disseminate retrieved security information to the member nations and, by so doing, engender increased confidence in the evolving world system. And, because such an enterprise would require major cooperation and exchange, it would likely simultaneously enhance international communication and understanding, which, in turn, could further contribute to increased international stability.

Seismological Verification

An international network of seismological stations could help to verify compliance with nuclear test bans, comprehensive and otherwise. In 1976 in Geneva, the United Nations Committee on Disarmament proposed a set of guidelines that could be used to implement a program of cooperation among national seismological facilities.[15] In such a system, which would benefit nations that do not have the resources to accomplish the task, central offices would interpret the compiled data and prepare comprehensive reports so that judgments about the causes of different disturbances could then be made with confidence and in unanimity. Combined with satellite verification, such a network could make it extremely difficult to test nuclear weapons without alerting the international community.

Radar Systems

In a nuclear-weapons-free world, radar systems, like surveillance satellites, could be used to assist the goals of anticipation and prevention. International peacekeeping forces as well as national military establishments could benefit from such technology by obtaining early warning of potential aggression. Examples of modern developments in this field include sophisticated surveillance systems mounted on reconnaissance aircraft, typified by the US Joint Surveillance Target Attack Radar System (JUSTARS), and over-the-horizon radar (OTH), which is diffracted and scattered by the earth's atmosphere to achieve greater range than conventional systems.

Submarine-Tracking Technology

Technologies capable of locating and tracking submarines also can contribute to the overall security of a nuclear-weapons-free world. In addition to sonar, a number of technologies have been devised or are under investigation for this purpose, including acoustical surveillance, the monitoring of heat emissions, the use of lasers, the detection of a submarine's magnetic field, and the observation of surface or interior waves.[16] Although the feasibility of a reliable submarine detection system is a matter of scientific dispute, the development of such a system could be another step in providing security against submarines that carry conventional weapons or, in the case of treaty violations, nuclear weapons.

High-Tech Conventional Defense

Even in a radically altered global system, free from the current nuclear standoff, the problem of disparities in conventional armed force structures will persist. The development of sophisticated *defensive* weapons, however, in the form of, for example, remotely piloted vehicles (RPVs) and precision-guided munitions (PGMs) – *i.e.*, accurate, target-seeking weapons launched from ground-based artillery systems or aircraft – could help to reduce or even eliminate the consequences of such imbalances. The potential advantage of such weapons and weapons systems is that, when integrated into an intelligent defensive scheme, they make conventional invasion difficult or impossible and are, in addition, far less expensive than the weapons they are designed to destroy, *i.e.*, armored vehicles, aircraft, and surface ships. Several plans of this type already have been suggested for the conventional defense of Western Europe.

SECURITY DEPENDS ON MORE THAN TECHNOLOGY

Of course, the ultimate utility of the foregoing and like technologies in helping to safeguard a post-nuclear world necessarily will depend on basic changes in other dimensions of the global system. Without corresponding reforms or transformations in the socioeconomic, political, legal, and other realms, technological answers to the national and international security problem are more likely to perpetuate rather than assist the struggle for progressive change. Without an overriding geopolitical regime whose endpoint is the termination of arms-racing once and for all, the selfsame technologies could prove catastrophically destabilizing. For example, stability enhanced by mutual satellite surveillance could turn into its opposite in a time of resumed cold war crisis. Under such circumstances, satellite observations otherwise satisfactorily explicable or not seen at all with lesser technology could be interpreted as threatening, leading to responses that likewise could be interpreted as threatening, thus creating an ominous self-fulfilling run-away situation compounding rather than allaying fear and suspicion of imminent danger. Remotely piloted vehicles and precision-guided munitions, while holding the line in times of trust, can as well be used with devastating consequence in times of war. One has but to reflect on the role of the machine gun or the introduction of aircraft into World War I to appreciate the downside of nearly every alternative security technology.[17]

Also, it bears notice that the technologies briefly noted here derive from a definition of security that emphasizes the avoidance of military confrontation and conflict. A broader conception, informed by a desire to avoid mounting ecological and economic threats, would result, of course, in a wider array of technological recommendations aimed at thwarting environmental decay, stimulating economic well-being, and otherwise enhancing the human condition. Form follows function.

But considerations such as these are beyond the primary purposes of this chapter, which, in addition to recommending some technological options to help safeguard a post-nuclear world, are to demonstrate the inherent limits of technology relative to security and to expose the highly questionable ways in which technology-security decisions have heretofore been made, with a view to suggesting an alternative to the nuclear threat system upon which, to our enormous *insecurity*, we currently rely. It is my fundamental belief that no amount of technological innovation—least of all such assumed marvels as the Strategic Defense Initiative or SDI, which I prefer to call the "Self-Deception Initiative"—can substitute for informed, humane, and rational decision-making in the quest for lasting security. There exists a great tendency to engage the national and international security debate at the level of specific weapons and technologies. Interesting as such debates may be, however, they forestall serious consideration of the ways in which the application of technology to security has been perverted and, as well, the ways in which technology can be made to serve the fundamental socioeconomic and political bases of true security so as to benefit all humankind.

THE MISAPPLICATION OF TECHNOLOGY TO SECURITY

It is a timeless maxim that we can use technology either to promote the common good or to pursue shortsighted, self-serving interests. Technology itself has no morality; it is only the *applications* of technology, arising as they do from human choices, that can be judged by moral standards. Technology, and the opportunities it affords, do not arise spontaneously out of a vacuum. Humans create technology, and, generally, other humans decide its application. An implicit covenant exists, therefore, between those who create technology and those who apply it.[18]

Unfortunately, humankind has done violence to this covenant, especially when it comes to applying technology to security. Instead of promoting the common good, we too often have misapplied or squandered the precious opportunities that technological advances have made available to us. The history of US national security policy at Hiroshima and Nagasaki during World War II and its conduct in the arms race in the years since demonstrate a generally lamentable quality of decision-making by human beings faced with choices laid open by technology.

Hiroshima-Nagasaki

The US decision to use atomic weapons against Japan in 1945 may be said to have been among the most seriously flawed and regrettable decisions ever made in human history. President Franklin D. Roosevelt approved the Manhattan Project in 1942 to develop an atomic bomb in response to rumors that Nazi Germany already had such a project under way. That is, the US effort was conceived as an attempt to meet a perceived *German* threat (although, as it turned out, the Germans had decided to "scuttle the project to develop an atom bomb" and instead opted to develop "an energy-producing uranium motor for propelling machinery"[19]). Relative to *Japan*, however, the facts suggest that far less honorable and altruistic motives were involved.

In the first place, no claim ever has been made that Japan was developing atomic weapons also. Instead, US officials and others have justified the Hiroshima-Nagasaki bombings primarily on the grounds that upwards of 500,000 to 1 million US deaths were averted. In his memoirs, President Harry S. Truman claimed that the bombings saved perhaps 500,000 US combat lives.[20] Secretary of War Henry L. Stimson similarly conjectured "over a million casualties, to American forces alone."[21] And British Prime Minister Winston Churchill claimed in his memoirs that a failure to use atomic weapons "might well require the loss of a million American lives and half the number of British. . . ."[22] Most Americans have accepted this "mythinformation" without question. But credible conflicting evidence strongly suggests that the figures used were more self-serving then real. In mid-June 1945, the US Joint Chiefs of Staff had prepared a plan for the invasion of the southern Japanese island of Kyushu; and, contrary to the figures later popularized by Truman, Stimson, and Churchill, their worst-case estimate of US deaths from the planned invasion was not more than 20,000 and probably less than 15,000 (an estimate that presumed 5,000 air and naval losses prior to the invasion, no more than 10,000 during the invasion itself, and 5,000 unforeseen collateral losses). Indeed, it is difficult to imagine that the Joint Chiefs of Staff would ever have recommended the Japanese invasion plan to the President had the estimates actually been anything like those quoted after the fact.[23]

Also noteworthy is the fact that almost a full month before the fateful *Enola Gay* flew over Hiroshima, the Japanese were making aggressive peace overtures through the Soviet Union, which had entered the war against Japan at the urgings of Great Britain and the United States at Yalta in February 1945 and had subsequently amassed troops in Manchuria for the purpose of taking over South Sakhalin and the northern Japanese Kurile Islands, as secretly agreed to at Yalta. As early as July 12, 1945, four days prior to the successful Trinity test at Alamogordo, New Mexico, the Japanese Emperor had requested Foreign Minister Shigenori Togo to instruct Japanese Ambassador Naotake Sato in Moscow to inform Soviet Foreign Minister Vyacheslav M. Molotov "that the Emperor wanted the war ended immediately and wished to send Prince Fuminare Konoye to Moscow with power to negotiate a peace on almost any terms, presumably short of the unacceptable sacrifice of the imperial dynasty."[24] These facts were known to the United States on July 13 as a result of intercepted and decoded cables and were carried by Secretary of the Navy James V. Forrestal to President Truman and Secretary of State James F. Byrnes, then meeting with Churchill and Soviet Premier Joseph Stalin at Potsdam. The gravity of the Japanese situation and, hence, the credibility of the surrender overtures are suggested by the precipitous decrease of US aircraft losses from bombing missions over Japan, from a high of 5.7 percent in January 1945 to 0.4 percent by July.[25] Because   Japanese   air defenses had been so compromised, the United States, by the end of May, had bombed virtually every major city in Japan – except, be it noted, Hiroshima and Nagasaki, which, in spite of their accessibility to US bombers, were spared conventional bombing. On the night of March 9 alone, more than 100,000 men, women, and children were killed in the incendiary fire-bombing of Tokyo.[26]

Finally, it is important to note that the bomb detonated over Hiroshima on August 6, 1945, producing 68,000 immediate fatalities and 76,000 injured, was an untested gun-type bomb dropped only twenty days after the first successful test of an atomic implosion-type bomb at Alamogordo, New Mexico, on July 16.[27] The Japanese peace overtures, which began prior to the successful Trinity test and which were known to the United States well before August 6, were of no avail; likewise the voices that were raised in a petition to the President circulated by Leo Szilard among his fellow atomic scientists in July 1945, foreswearing the use of atomic weapons against innocent non-combatants. These efforts fell on deaf ears, as did also proposals to demonstrate the power of the bomb before Japanese civilian and military officials prior to its actual use. The stumbling block was President Truman's insistence, in the face of considered recommendations to the contrary from his own staff, on unconditional surrender – that is, conditions prohibiting retention of the revered Japanese emperor. In the end, after 38,000 more immediate fatalities and 21,000 more injured at Nagasaki, the Emperor was retained. A weapon conceived in response to a technologically superior German war machine was, instead, used against essentially defenseless, non-combatant Japanese civilians.

It is evident, then, that the rush to use atomic weapons against Japan had less to do with legitimate US military objectives than it did with denying the Soviet Union access to that country. Even then,

fearing lost opportunity and despite ongoing attempts on the part of
the Japanese to obtain Soviet mediation of terms of surrender, 1.6
million of Stalin's troops attacked Japan from their Manchurian
encampments on August 9. The 106,000 dead and 97,000 injured in
the opening salvo of the nuclear era thus should be tabulated as the
first victims of the Cold War between the United States and the
Soviet Union. Never in the history of modern warfare had a major
new weapon been taken from development into combat in a matter
of a few days, let alone one not tested before combat use, so
confident were its designers that it would work on the first try.[28]
Disquieting also is the sparing of the cities of Hiroshima and
Nagasaki from conventional bombing in anticipation of the atomic
bomb, raising the specter of an experiment to determine the lethality
and specific effects of the new weapons in a pristine setting
uncompromised by prior weapons damage. If, as some have argued,
unacceptable casualties had to be inflicted upon civilians to force an
early surrender, the evidence from Tokyo makes it clear that the
same effect could have been accomplished without resorting to
atomic weapons.[29]

In light of the evidence, the use of atomic weapons against
Japan was a reprehensible and cowardly act. The first use of atomic
weapons by the United States represented what Manhattan Project
physicist Leo Szilard referred to as "a flagrant violation of our own
moral standards." It may be said also to have constituted a violation
of international law[30] and, perhaps most importantly, a violation of
the implicit trust between political leadership and the US people,
including those whose genius actually created the weapons. But the
saga of the arms race is rife with examples of faulty logic and
dubious reasoning to justify the misappropriation of technology in
the name of national security.

Arms Control

In the field of arms control, for example, the United States has
seen itself as committed in principle to the abolition of a system in
which (presumed) security depends so heavily on technologically
sophisticated weaponry, but has demonstrated itself to be perpetually
short of the will to take the first really significant steps mandated by
that principle. Indeed, driven by fear verging at times on hysteria
and by such other forces as economic greed and the lust for power,
the United States has come to measure security in megatons and,
consequently, to countenance a long series of negotiated settlements

that, by design, have left humanity more imperiled than when the process was entered into in the first place. Whether or not the current reconstitution of Eastern Europe will reverse or retard this practice remains to be seen. But if past behavior is any indication of future conduct we should not be overly optimistic. History shows that "bargaining chips" created by US weapons technology never have been used as such – that is, weapons have never been discarded in exchange for a significant arms control objective.

A remarkable opportunity to end the arms race once and for all, largely unknown to the US public, presented itself in 1955 when the United States enjoyed an overwhelming strategic superiority over the Soviet Union and consequently could deal with the Soviets from the much-coveted "position of strength."[31] In December 1954, a joint British-French proposal was put before the United Nations Disarmament Commission calling for a treaty that would lead to complete nuclear disarmament under international control. The proposal was endorsed by all the Western States, including the United States, with the Soviet Union being the only holdout. In Spring 1955, however, in a move that is not without precedent, the Soviets presented their own proposal incorporating all the main features of the original British-French plan, and the United States responded favorably.[32] Finally, the nuclear powers had agreed upon a formula to end the arms race. The arms race was to be called off.

But the air of optimism and elation was short lived. On September 6, in a dramatic reversal of the US position, the new US delegate to the Commission, Harold Stassen, announced that the United States "put a reservation" on all of the substantive positions it had previously taken in the disarmament commission or at the UN on questions relating to levels of armament.[33] Though then President Dwight D. Eisenhower represented himself publicly as eager to reduce the threat of nuclear war, it is evident that he was caught off guard by the unexpected Soviet acceptance of the terms and conditions officially endorsed by the United States. Embarrassing as the sudden reversal must have been, it was preferable to suffer the temporary criticism that would follow than to be drawn into a policy that would force the United States eventually to abandon its nuclear supremacy. That supremacy, history has shown, was to be eroded eventually, precisely because the arms race was allowed to continue. However, to have seen this possibility and acted accordingly would have required a modicum of vision and statesmanship not generally forthcoming from individuals whose political survival is wedded to the cultivation of the fear of an external enemy. The initial US endorsement was, evidently, disingenuous, and one of the greatest dividends ever yielded on our technological investments was thrown recklessly aside.

Another example of the misappropriation of weapons technology to national security began in 1970 under the Nixon Administration. It has had repercussions to this day, leading, as I will explain, to the "Star Wars" proposal of President Reagan.

By Spring 1970, US weapons technology had advanced to the point of developing the capability of arming individual strategic missiles (launchers) with multiple independently targetable reentry vehicles (MIRVs). Prior to that time, US and Soviet land-based strategic forces formed a stable mutual deterrent because the missiles on each side carried only one reentry vehicle (warhead) per missile – a situation of mutual stability that prevailed for many years, with neither side motivated to alter the basic configuration of its land-based forces.[34] The MIRV option, however, was inherently destabilizing. By launching several warheads from each missile, one side could hope to destroy all of the other's launchers and still have missiles to spare for other contingencies. When it became known that the United States was moving to deploy its new MIRV technology on its land-based missiles, the Soviet Union, which did not then have this technology, objected strenuously on the grounds that stability would be lost and the risk of nuclear war thereby increased. At the SALT talks in Vienna in May 1970, the Soviet Union made twenty-six separate overtures to the United States to defer deployment and to enter instead into negotiations to ban MIRV deployment by either side altogether. The response of the Nixon Administration was to refuse to regard these overtures as serious, saying that it was "a trick designed to dupe the United States into delaying MIRV deployment."[35] In characteristic fashion the new technology, developed as a bargaining chip to get the Soviet Union to the negotiating table, was deployed; the bargaining chip was never used as such and a critical opportunity was lost. Having been rebuffed, the Soviets proceeded to develop their own MIRV capability in about five years' time.

But this was not the end of the story. Looking down the Soviet MIRV barrel, the United States saw a "window of vulnerability" for its own land-based strategic missiles. Failing to recognize that the problem was of its own making and failing to learn from past mistakes, the United States looked for yet another "technological fix," which soon came in the form of the MX missile proposal. To be shuttled in mobile protective shelters (MPS) along thousands of miles of railroad tracks, the MX was designed to slam shut the alleged window of vulnerability on the US side. When, however, it was recognized that the MX would be the largest human project ever, that it would be extremely expensive, that it would create a national shortage of concrete and have major negative ecological effects, and that, despite its MPS strategy, it still would be a prime target for Soviet missiles, the selling of the project began to bog down.

Enter the Scowcroft Commission, set up by President Reagan to save the MX missile proposal (and named after its chairman, Lieutenant General Brent Scowcroft, currently National Security Advisor to President Bush).[36] Using arguments put forth in vain for years by analysts out of phase with Washington's needs,[37] the Commission came to the conclusion that the window of vulnerability never existed, that the MPS idea should be scrapped, and that the MX missiles should be dropped into existing Minuteman missile silos. The desired effect was achieved: The day was saved for the MX missile and Congress dutifully appropriated the requisite funding. No one asked whether the very concept of land-basing strategic missiles might have been made obsolete by the US and Soviet MIRV decisions. No one asked whether the United States might get a leg up on the Soviet Union simply by abandoning land-based missiles in favor of relatively invulnerable strategic submarines and bombers. No one asked whether the Soviets, who had just completed a costly land-based missile modernization program and who had 73 percent of their strategic warheads on land compared to 21 percent for the United States, might be in a most unenviable position. And certainly no one seemed to notice or object to the circular logic involved: The MX proposal was inspired by an alleged window of vulnerability, yet when that window was declared non-existent by credible authorities, the MX proposal was not consigned to the dust bin.

Of course, unprotected human beings could not be made invulnerable by the fiat of a presidential commission, and so President Reagan proposed to create, in the form of SDI, an impenetrable shield over the United States to protect the US population against nuclear missile attack. The Office of Technology Assessment and all but a cadre of diehards quickly recognized that a general population protection scheme of this sort was not possible for the foreseeable future and probably never would be. But rather than abandon the idea as a pipe dream fraught with major technological problems, Washington took a fall-back position: If SDI would be too leaky to protect human populations, at least it could provide a partial defense for our land-based missiles, thereby enhancing their deterrent value. In this context, naturally, the Scowcroft Commission's finding that no window of vulnerability existed was not to be mentioned too loudly. A number of other observations were also downplayed, such as the possible obsolescence of land-based strategic missiles and the commensurate desirability of abandoning allegedly vulnerable land-based strategic missiles in favor of relatively invulnerable strategic submarines and bombers. SDI was to be deployed to protect land-based missiles, including the MX, which itself was deployed to close an alleged window of vulnerability opened by the earlier failure of the United States to relinquish deployment of MIRV technology. Put another way, if MIRV had not been deployed and instead had been banned by both sides, there would be no more justification for "Star Wars" today than there was at any time after the first installation of land-based missiles and prior to the MIRV decision in 1970. Once again, near-term political ambitions prevailed over long-term considerations of national and international security.

Such circumscribed thinking appears to continue even in the face of a world changed almost beyond recognition as a result of the sweeping reforms introduced by Mikhail Gorbachev, reforms not dreamed of since V.I. Lenin, and now spreading throughout Eastern Europe. Concessions that for the entire course of the arms race have been "pie in the sky" are being granted unilaterally, and proposals that go beyond even our wildest imaginations are being tendered seriously. The Soviets have taken the moral high ground by dint of the US failure to seize prior opportunities opened by its consistently superior strategic technology. Yet, in spite of it all, every major weapons system put into the pipeline in the first Reagan Administration – the resurrected B-1 bomber, the Trident II missile, the MX missile, SDI, massive procurements of cruise missiles, the 600 ship Navy, 17,000 new nuclear warheads, the advanced technology (Stealth) bomber and Stealth cruise missile, a massive commitment to C^3I (command, control, communications, and intelligence) – is still in the pipeline or reaching deployment. There is, I believe, essentially no relationship between reality and the conduct of the arms race, exposing in a dramatic way the specious quality of most of the arguments proffered in its behalf and, once again, the misapplication of technology to security. When ordinary mortals behave in ways unrelated to the reality around them, psychiatrists call them insane.

Of course, a major explanation for this craziness is to be found in the cast of characters participating in the US application of technology to security, which extends far beyond the politicians and bureaucrats reported daily by the print and broadcast media. Defense contractors and government laboratories, whose interests do not necessarily comport with the greater good, are major players in the arms race, and they enjoy altogether too much influence in the process.[38]

This point is made abundantly clear by the history of the decision to manufacture and deploy the Pershing II missile. Under development by the Martin Marietta Corporation since 1974 as a modernization of the Pershing IA already deployed in Europe, it was to have a more accurate upgraded warhead but otherwise the same short range as the Pershing IA, precluding targeting the Soviet Union from Western Europe. However, early in 1978 Martin Marietta officials became aware that long-range weapons were under discussion in the secret meetings of a special nuclear committee of the NATO ministers known as the "High Level Group" (HLG). Moving quickly, Martin Marietta proposed directly to US Army officials and members of the HLG that a second-stage rocket be added to the missile as part of its upgrade, enabling it easily to strike targets within the Soviet Union. The marketing maneuver succeeded. Persuaded that the addition of a second-stage rocket would be largely overlooked in what otherwise would be perceived as a routine upgrade, the NATO ministers accepted the Martin Marietta proposal and left it to the United States to determine the precise number of missiles to be deployed and the types of warheads they were to carry. Stated Hans Apel, then the West German defense minister, "we didn't anticipate any problems. It looked so simple just to replace the Pershing IAs with more modern weapons. Nobody thought about any political repercussions."[39] Though promoted subsequently in response to the deployment of the Soviet SS-20 missile (itself a modernization of the aging Soviet SS-4 and SS-5 missiles), the Pershing II that finally emerged was more the result of a marketing ploy than a considered response to a perceived military need – an application of technology to security matters driven by marketing strategies intended to enhance the portfolios of the stockholders of a defense contractor rather than the long-term security of national and international society.

Clearly, such maneuvers raise serious ethical questions. What is the propriety of defense contractors having direct access, for marketing and promotional purposes, to high-level NATO ministers?

By what right do company officials gain access to the secret deliberations of committees such as the HLG? What is the role of the US government in establishing and promoting such arrangements? These and related questions are of the utmost gravity and relate to the *de facto* surrender of strategic and foreign policy-making to unnamed business persons well beyond the influence or scrutiny of the US people.

"Decapitation" Scenarios

On October 13, 1981, the opening session of the first of an annual series of unclassified National Security Issues Symposia, designed to present and explain the Reagan Administration's national security policy to a responsible cross section of the lay community, was held in Bedford, Massachusetts. That day, Dr. Richard Pipes, a staff member of the National Security Council, presented a candid assessment of the Administration's view of the Soviet Union. He stressed, in particular, those factors in the development of the Administration's strategic nuclear policy that in his judgment justifiably distinguished it from all previous policies, to wit, that it was the first to comprehend fully the deterrent value of "counterforce" strategy.[40] Prior administrations, Pipes maintained, had mistakenly judged the Soviets to value life and property as we do and therefore had naively implemented predominantly "countervalue" strategies threatening life and property. The Soviets, he contended, value not life and property but, rather, their weapons, their communications and control systems, and their political system. Therefore—and this was the major breakthrough of the Reagan Administration according to Pipes—deterrence was best served by so-called decapitation scenarios, in which the Soviet command and control structure is the preferred target.[41]

Any high school student can recognize immediately not only the cultural hubris, but, as well, the fatal flaw in the logic of these remarks: You can't talk to someone you've just decapitated. Nations and their leaders do not exist, like Schroedinger's cat, in some kind of quantum mechanical "superposition of all possible states" that would enable one to communicate with the dead.[42] Later the same day, Lieutenant General Brent Scowcroft, then an Administration consultant on foreign policy and national security affairs, wrestled with the implications of decapitation scenarios:

There's a real dilemma here that we haven't sorted out. The kinds of controlled nuclear options to which we're moving presume communication with the Soviet Union, and yet from a military point of view one of the most efficient kinds of attack is against leadership and command and control systems. Much easier than trying to take out each and every bit of his offensive forces. *This is a dilemma that I think we still have not completely come to grips with.*[43]

Indeed! The dilemma fairly screams out that something is dreadfully wrong with our logic. Yet the power of ideology over rational thought is so great that evidently few seem to have spotted the fly in the ointment. As viscerally appealing as their advocates may find them, decapitation scenarios make no sense whatsoever. Should a decapitation ever be carried out, the most likely response would be an uncoordinated paroxysmal release of virtually every strategic weapon in the opposition's arsenal, extinguishing essentially all life on earth.

This indictment of the fundamental illogic that lies at the core of US strategic policy is reinforced by what I infer from an exchange between General Scowcroft and Dr. Edward Teller (former director of the Lawrence Livermore laboratory and considered by some to be the father of the hydrogen bomb) at the National Security Issues Symposium in Bedford, Massachusetts, in 1981. Dr. Teller asked two questions of Scowcroft, the second of which was as follows:

There is a second question which I'm afraid we cannot, for many reasons, discuss in detail. But I would at least like to mention it. Has it been considered to concentrate the power of the President and his various successors [in the event of decapitation by the Soviets] on various levels not only on commanding an attack – a counterattack – but on restraining the usefulness of the weapons in such a manner that this restraining power automatically lapses unless renewed periodically? And it also lapses automatically on the unmistakable signs of actual explosives hitting the United States? What I have in mind is a hierarchy where it does not pay the enemy to attack the top leadership because all they do thereby is lose the peoples with whom they can negotiate; and instead . . . they have an interest in preserving our leadership. Has that been considered?[44]

Replied General Scowcroft: "You are very correct Dr. Teller, we can't talk about it. Is that enough of an answer?"[45] To which Teller responded: "A little more than enough. Perhaps I have to apologize about mentioning that such an idea could even exist."[46]

My reading of this exchange is that the United States very likely has in place some automatic retaliatory response to be activated in the event of a confirmed attack on US soil or, most certainly, a confirmed loss of the US National Command Authorities.[47] If so, the astonishing conclusion is that, if the much-trumpeted deterrent value of our force structures and strategies fails, the United States will, with malice aforethought, commit national suicide along with the massive collateral humanicide of most, if not all, of the rest of the world's population. This is, I believe, no less than an intent to commit a crime against humanity of the highest order and it therefore merits the strongest possible condemnation. That any nation even entertains such an idea should be a matter subject to the jurisdiction of the International Court of Justice. Yet, to my best knowledge, this exchange, which took place in the presence of reporters from at least two major US newspapers and later was published in the official transcript of the National Security Issues 1981 Symposium, never was reported to the US people, never was brought up before Congress, and certainly never became a subject of national debate. Besides being a remarkable example of the selective inattention of individuals of all persuasion, however, it demonstrates again the extreme misapplication of the technologically feasible in the name of national security.

At a minimum, the exchange lends credence to the proposition that we are moving closer and closer to automated responses that may function after our demise and that do not serve the true national interests of the living. As such, it raises a host of troublesome questions. What if the attack on the United States comes from a country other than the Soviet Union? Bearing in mind that it takes weeks—even months—to sort out the precise causal sequence of events in disasters far less significant than a nuclear war, might we unleash the full fury of our nuclear arsenals upon the Soviet Union in response to an attack from a Third World country? How discriminating is an automated response under fire likely to be? And what is the likely Soviet response to the knowledge or suspicion that we have such a plan? When pressed on the logic of targeting Soviet political and military command and control centers, many, including General Scowcroft, have stated their belief or hope that the Soviets would not respond paroxysmally if "decapitated." And perhaps, today, in the glow of *glasnost* and *perestroika*, there is some justification for this belief or hope. But if the United States, the alleged "good guys," have already contemplated such a plan, who can be sure that the Soviets, not long ago branded the "focus of evil" in the world, will not do likewise? The quality of thinking here is manifestly deplorable!

The Reagan Administration's analysis, as articulated by Dr. Pipes, is a textbook example of the technique of deprecating the opposition to justify the undertaking of otherwise unwarranted decisions and actions. In this case, the allegation, accepted *carte blanche*, that the Soviets "eat babies" was used to justify and motivate the massive arms buildup initiated in the first Reagan term. No one questioned seriously whether Dr. Pipes really knew what he was talking about; whether he or others might not have had a hidden agenda; whether the claims about the Soviets would stand the test of critical examination; and, in any case, whether the Administration's proposed responses were in any way appropriate or beneficial. The less reasoned the exhortation, it seems, the less reasoned the response.

TOWARD THE INFORMED, HUMANE, AND RATIONAL APPLICATION OF TECHNOLOGY TO SECURITY

In the preceding section I identified and illustrated some of what I believe to be the perverse ways in which, beginning with Hiroshima-Nagasaki, technology has been (mis)applied in the name of national and international security. Space permitting, each could be supported by many more illustrations and analyzed in far greater detail, and the areas of principal concern themselves could be extended considerably beyond those I have chosen. However, the material I have presented is already sufficient to identify unambiguously the broadest and most fundamental reforms that are required if we ever are to find lasting and stable security without relying on nuclear deterrence. Our hope cannot continue to reside with an indefinite series of technological fixes that become exponentially ever more complex and bizarre.

The first necessity and, unfortunately, the one that will be far and away the most difficult to achieve, is a revolution in the way people think of themselves and their country; in their general level of interest and knowledge of domestic and foreign policy; in their tolerance of alternate viewpoints both from within and without; and, above all, in their willingness and commitment to be faithful to notions of human solidarity. Americans, for example, should take the foreigners' image of the "ugly American" seriously and examine it for its true meaning and content. Everyone can learn much by listening carefully to their detractors. In the final analysis, genuine security will be derived from the genuine respect—not fear—of others.

Of course, human nature being what it is, this sort of revolution cannot be legislated into existence. We are talking about the way people think, and that is ordinarily very much dependent on the way we think those around us think. Herein lies the power of jingoism. People are exhorted to think uniformly and to derive their sense of the correctness of their thoughts from the reinforcement of those around them, rather than from self-examination and reflection. The latter traits are not seen as virtues and those exhibiting them are judged to be out of the mainstream, a place where, presumably, no one wants to be. Therefore, if societal thinking is to change in the fundamental way I believe it must, it probably will require an enlightened charismatic national leader. Though he still must stand the test of time, Mikhail Gorbachev may well prove to be such a leader.

Of the areas that are amenable to legislative action, the most wanting is the need to institutionalize, with an eye to clearly defined objectives, the process whereby technology-security decisions are made. First, governments must deliberately insulate the making of strategic policy from the vagaries of political fashion. Second, they must assure that a comprehensive cross section of thought—political, legal, moral, technical—is brought to bear without prejudice upon all policy decisions. The process must be both flexible, in the sense of admitting the greatest possible range of analysis, and rigid, in the sense of being essentially immune to political caprice. Policy must square with the highest moral principles, withstand the most rigorous legal analysis, and be feasible in the technical sense, both short- and long-term, with a full accounting for all conceivable contingencies both favorable and unfavorable. Strong and effective measures must be instituted to minimize to the greatest degree possible the influence of parochialism and the squalid divisiveness of such base tactics as the dehumanization of other peoples.

The standard called for transcends any that can reasonably be expected of an elected body or official. I have in mind the establishment of national strategic policy-making bodies that are as free from political influence as, say, the judicial branch of the US government, but which, unlike the US judiciary, play an active rather than a passive role. Policy, and its realization, must be stable over the long term and must be consistently the product of the best thinking that can be brought to bear. There must be virtually no possibility that debacles such as MIRV and its chain of foreseeable consequences can come to pass. Likewise, exhortations that dehumanize the enemy and intellectual nonsense such as the decapitation conundrum never must see the light of day. In the United States and elsewhere, the executive and legislative branches of government, including the defense establishments and relevant others, must be obliged to carry out policy established by such bodies with failure to do so being regarded as a transgression of the highest order.[48] Through such bodies, the quality and consistency of strategic policies, as well as their implementation, can be elevated to levels commensurate with the complexity and danger of the problems the world is likely to face in the foreseeable future.

Obviously many lower level and, hence, more specific, reforms are called for as well. The first of these should probably be a sweeping reform of the way we solicit, develop, and procure weapons technology. As suggested by the example of the Pershing II missile, there must be strict laws limiting access of individuals and corporations with vested interests to policy-makers and their deliberations. Penalties for violations should be severe.

We need also to treat weapons-making for what it is: the lamentable diversion of human talent into the creation of sophisticated means to kill and mutilate fellow human beings. However this pastime may be justified, its moral implications must never be allowed to drift far from view. Weapons-making, if it must be, should be seen as a public duty, not as a glamorous high profit enterprise to which business-as-usual ethics and practices apply. To this end, defense contracting should, in effect, be nationalized. Weapons should be made only because it is deemed that they must be made, not because a company or corporation has a stake in a follow-up contract. To the greatest degree possible, the incentives to keep new weapons systems coming for their own – and a company's -- sake should be eliminated.

Defense employment should be deglamorized for individuals as well. Salaries should be set by law somewhat below comparable positions in civilian industry.[49] This would provide immediate relief to the civilian sector in terms of competitive access to the best engineering and scientific talent (which, in the case of the United States, might also mitigate somewhat the decline of US technology in world markets). It also would have a desirable sobering effect upon persons considering careers in the defense industry. It is to be assumed that as a result of the absence of nuclear weapons and the presence of policy-making reforms there would be a decrease in demand for weapons technology. If, in spite of this, diminished individual incentives lead to a shortfall of available workers, defense employment could be made a form of obligatory national service. After all, if defense employment really is for the national good, and not merely for high profit, high salary, and high glory, then the patriotic motive should suffice.

Predictably, proposals to take such a heavy hand to the defense industry and defense employment will be met with resistance because such actions are assumed to create economic dislocations that can throw tens or hundreds of thousands of people out of work. But this response — often morally reinforced by calls to sacrifice for the greater good, notwithstanding that the lives spared by an averted war tomorrow surely are worth the loss of jobs today — overlooks the fact that the claim of economic dislocation is without merit in the first place. As Professor Lloyd J. Dumas has pointed out with eloquence, defense employment is essentially a form of welfare.[50] The products of defense employment add almost nothing of value to the economy, where value is measured not in terms of dollars that have changed hands but in terms of goods and services that lead *directly* to the creation of additional goods, services, and employment. To use a familiar analogy, dollars spent on defense are like paying to have a hole dug in the ground and then paying to have it filled in again; money changes hands, but no real value in terms of the creation of new goods and products or meaningful employment is added to the economy. When measured in this way, there is very little difference between paying a defense worker to do defense work and paying the same worker to stay at home; if one simply continues to pay dislocated defense workers their salaries, as if they were on welfare, society would not be worse off economically than it is right now. On the other hand, if instead, as Seymour Melman has suggested, we were to invest a moderate amount in retraining dislocated defense workers for civilian employment, we could be a good deal better off than we are right now.[51] The talent freed up and properly retrained would constitute an enormous resource that could be applied to the revitalization of our economy and our industrial infrastructure, as well as to reversing the deterioration of health services, education, housing, roads, public transportation, the environment, our inner cities, and other social services and programs all of which have been compromised by defense spending. Such a conversion of resources could lead eventually to enhanced national security, not of the kind guaranteed at the point of a gun, but that serves as an example for the world to emulate.

Finally, I recommend a relatively minor reform that happens to be near and dear to me as a taxpayer and citizen concerned with the state of higher education in the United States. It is now the law in the United States that, to qualify for US government educational loans, male students must register for the draft. This law, which incidentally illuminates well our current priorities, should be repealed forthwith. In its place should be substituted a regulation requiring colleges and universities whose faculties receive technological and related research grants from agencies of the Department of Defense (whether or not the projects funded are classified) to provide mandatory standardized courses on the history and dynamics of the arms race to all degree candidates in fields of interest to the defense industry. It is unconscionable that young people whose talents are turned to the perfection of weapons and weapons-related systems should be largely ignorant of the political and foreign policy implications of their work. No world, whether free of nuclear weapons or not, can be genuinely secure when the generation that inevitably will someday take over the reins of power is ignorant in these respects.

CONCLUSION

The terms *glasnost* and *perestroika* and the unprecedented changes taking place in Eastern Europe in their name afford a literally undreamed of opportunity to apply the principles of alternative security to the now conceivable prospect of a world essentially free of nuclear weapons. In the spirit of the ancient Chinese curse, "May you live in interesting times," the blessing could devolve into tragedy if we fail to set aside our former ways and embrace the opportunity from all sides. This could well be our last chance to say no to the gut impulse to seek security through technological means. As I argued at the outset, we have advanced the scope and sophistication of technological wizardry to the point of vanishing returns, a condition from which there is no reprieve for reasons that are fundamental and beyond the ability of the human will to mitigate. That this opportunity should have presented itself at this propitious moment is worthy of the accolade "miracle." It would be prudent to conduct our affairs accordingly. Fortune is not likely to look so favorably upon us more than once.

NOTES

1. President Reagan made public his proposal for an impenetrable missile defense in a nationally televised speech from the Oval Office at the White House on March 23, 1983. The complete text is contained in *Presidential Documents* 19, 12 (Monday, March 23, 1983). The concept, subsequently referred to as the "Strategic Defense Initiative," or SDI, was immediately dubbed the "Star Wars" proposal.

2. Alice Kimball Smith and Charles Weiner (eds.), *Robert Oppenheimer Letters and Recollections* (Cambridge, MA: Harvard University Press, 1980): 293-94, *quoted in* Kurt Gottfried, "Technological Development, the Military Balance, and Arms Control," in Wolfram F. Hanreider (ed.), *Technology, Strategy, and Arms Control* (Boulder, CO: Westview, 1986): 126-27.

3. Because each launch vehicle may contain tens of warheads along with hundreds of decoys and other countermeasures such as radar-confusing chaff and aerosols, it is of the utmost importance to eliminate as many missiles as possible during their boost phase, before these elements are disbursed above the atmosphere. *See, e.g.*, Clarence A. Robinson, Jr., "Panel Urges Boost-Phase Intercepts," in *Aviation Week & Space Technology* (December 5, 1983): 50. Current missile designs have boost burn times in the range of approximately 180 seconds (US MX missile) to 300 seconds (Soviet SS-18 missile). Future missiles incorporating so-called fast-burn boosters could complete their powered boost phase in 60 seconds or less. *See* Ashton B. Carter, *Directed Energy Missile Defense in Space—A Background Paper*, U.S. Congress, Office of Technology Assessment, OTA-BP-ISC-26 (April 1984): Section 2.

4. In addition to the rationale as a ballistic missile defense (BMD), the original defensive role for SDI is now being supplanted also in favor of its much more plausible offensive anti-satellite capability. *See* William J. Broad, "Pentagon Plans an Offensive Role for 'Star Wars': Satellite Killers," *The New York Times* (November 27, 1988): 1.

5. Indeed, it is not inconceivable that the bulk of a war for which these capabilities are designed could be fought after the extinction of the beings that created them. It hardly seems necessary to emphasize that this is one of the reasons we have become progressively less secure as, in the name of enhanced security, we have emphasized greater reliance on technological means. We have pushed the arms race well into the realm of diminishing returns but steadfastly refuse to admit it, let alone take corrective action.

6. Forty million square miles is the approximate operational area from which the US Trident submarine can strike targets within the Soviet Union with warheads of the D5, or Trident II, missile.

7. For a discussion of keep-out zones and other defensive and offensive countermeasures, see *Anti-Satellite Weapons, Countermeasures, and Arms Control*, US Congress, Office of Technology Assessment, OTA-ISCV-281 (September 1985).

8. This concept, known as a "defended keep-out zone," might of itself

constitute a violation of and, hence, force withdrawal from, the 1967 Outer Space Treaty, the 1963 Partial Test Ban Treaty (14 U.S.T. 1313, T.I.A.S. No. 5433, 480 U.N.T.S. 43) and the 1972 Anti-Ballistic Missile Treaty (23 U.S.T. 3435, T.I.A.S. No. 7503), depending upon the technologies deployed and how they are tested.

9. In this context, "complex" refers to systems with large numbers of independent distributed components utilizing widely divergent technologies and operating in a wide range of environments. One might suggest, for instance, that modern computers are counter-examples of extremely complex systems for which the problems of system integration are solved routinely. However, computers are not complex in the sense intended here.

10. In fact, it is argued with respect to SDI that systems integration may never be feasible at a level that would assure an acceptable defense, even if solutions were found for the daunting array of technological challenges that remain exclusive of integration. The tasks subsumed under systems integration for SDI include the so-called battle management function in which computers are used to control individual weapons and to coordinate their operation. This aspect of systems integration alone may be impossible at an acceptable level of reliability. *See, e.g.*, Herbert Lin, "The Development of Software for Ballistic-Missile Defense," *Scientific American* (December 1985): 46-53.

11. For example, at the time of the first Apollo lunar landing, a good deal of the incorporated technology was at least a decade old notwithstanding the fact that the elapsed time between President Kennedy's announcement of the goal to put humans on the moon and the successful flight itself was less than a decade. The reason is that commitments were made to technologies already no longer the state of the art at the time of the announcement because relatively little was known about the newer technologies and the older technologies were considered more reliable. Though it was recognized that better, even possibly more reliable, methods might be available and proven downstream, they could not be considered without introducing delays that would take the first successful flight into the next decade; Kennedy had set the mission to be achieved before the end of the decade in which it was announced: the 1960s. Similarly, relative to the US space shuttle *Challenger* disaster in 1986, it has been observed that aspects of the shuttle design were nearly fifteen years out of date.

12. *See* Ralf Trapp, "Verification of an International Agreement Banning Chemical Weapons – The Possible Role of Satellite Monitoring," in Bhupendra Jasani and Toshibomi Sakata (eds.), *Satellites for Arms Control and Crisis Monitoring* (New York: Oxford University Press, 1987).

13. *See* Jasani and Sakata, *Satellites for Arms Control and Crisis Monitoring*.

14. *See* Statement of President Valéry Giscard d'Estaing, 10 U.N. GAOR Ad Hoc Committee (3d plen. mtg.), U.N. Doc. A/S-10/PV.3 (May 25, 1978): 39.

15. Ola Dahlman, "Verification in Arms Control and Disarmament

Treaties and Negotiations—Towards a European Satellite Verification System?" in Jasani and Sakata, *Satellites for Arms Control and Crisis Management*.

16. *See* Tom Stephanick, *Strategic Anti-Submarine Warfare and Naval Strategy* (Lexington, MA: Lexington Press, 1987).

17. Similarly, progress in submarine surveillance calculated to build confidence and security in a post-nuclear world could have quite the opposite effect should there be a breakdown of trust and the resumption of a cold war, potentially waged with a threatened "reinvention" of nuclear weapons. The assured invulnerability of major strategic assets, such as submarines and bombers, is absolutely essential to the maintenance of restraint in a time of cold war crisis, because only thus is a preemptive first strike avoided. But we could find ourselves in an incalculably worse position than at present if the fruits of well-intended efforts are not accompanied by the political and other reforms that could ensure that the alternative technologies would be used benignly.

18. I use the phrase "those who create technology" here in the most general sense to include scientists and technologists, as well as the citizens and taxpayers who underwrite the application of technology to weapons development.

19. Comments of Albert Speer, Minister of Armaments and War Production of the Third Reich, *quoted in* Richard Rhodes, *The Making of the Atomic Bomb* (New York: Simon and Shuster, 1986): 405.

20. Harry S. Truman, *Memoirs: Years of Decision* 1 (Garden City, NY: Doubleday, 1955): 417.

21. Henry L. Stimson, "The Decision to Use the Atomic Bomb," *Harper's Magazine* (February 1947).

22. Winston Churchill, *The Second World War: Triumph and Tragedy* 6 (Boston: Houghton Mifflin, 1953): 638-39.

23. The majority of conservative estimates placed the total anticipated deaths somewhere between 7,000 and 8,000. The inflated estimates, however, put the cost of the Japanese invasion at between 70 percent and 240 percent more than the total of 292,000 lives lost by the United States during all of World War II. *See* Rufus E. Miles, Jr., "Hiroshima: The Strange Myth of Half a Million American Lives Saved," *International Security* 10, 2 (Fall 1985): 121-40.

24. Miles, *Hiroshima*: 127.

25. *The War Reports of General of the Army George C. Marshall, General of the Army H. H. Arnold, and Fleet Admiral Ernest J. King* (Philadelphia: J. B. Lippincott, 1947): 439.

26. Rhodes, *Making of the Atomic Bomb*: 599.

27. *See* Samuel Glasstone and Philip J. Dolan (eds.), *The Effects of Nuclear Weapons* (United States Department of Defense and Energy Research and Development Administration, 3rd ed., 1977): Table 12.09 at 544.

28. This fact, incidentally, gives the lie to one of the objections raised

against a demonstration shot prior to combat use, namely, that such a shot ought not be risked for fear of embarrassment should it fail.

29. Whether an experiment involving the lives of hundreds of thousands of defenseless non-combatant human beings actually was performed, the appearance of such an experiment has contributed negatively to the image of the United States ever since.

30. *See, e.g.*, The Shimoda Case, Judgment of December 7, 1963, District Court of Tokyo, translated into English and reprinted in full in *Japanese Annual of International Law* 8 (1964): 212. *See also*, for pertinent and extensive discussion, Burns H. Weston, "Nuclear Weapons Versus International Law: A Contextual Reassessment," *McGill Law Journal* 28 (1983): 542.

31. In 1955 the United States had about 1,100 nuclear-capable B-47 and B-52 intercontinental bombers; the Soviets had approximately 300 Tu-4's (copies of the US B-29) of limited range. Moreover, the US then, as now, deployed its bombers from bases distributed around the world, including many relatively close to the Soviet Union. The Soviets then, as now, had no comparable bases. Intercontinental and submarine-launched ballistic missiles were not then part of the arsenals of either country. *See* Randall Forsberg, "A Bilateral Nuclear-Weapon Freeze," *Scientific American* (November 1982): 52-61. Similarly, between 1951 and 1955 the United States had conducted approximately 56 nuclear tests. In the same period the Soviets conducted 10 tests. *See* Herbert F. York, "The Great Test-Ban Debate," in *Progress in Arms Control?* (San Francisco: W. H. Freeman and Company, 1978): 17-25.

32. Within two days, the US delegate responded: "We have been gratified to find that the concepts which we have put forward over a considerable length of time, and which we have repeated many times during this past two months, have been accepted in large measure by the Soviet Union." *Verbatim Records of the Meetings of the Subcommittee of the United Nations Disarmament Commission, April 19-May 18, 1955*, U.N. No. 3, Cmd. 9650 (1956), *quoted in* Steve J. Heims, *John von Neumann and Norbert Wiener* (Cambridge, MA: MIT Press, 1980): 268.

33. *See* Heims, *John von Neumann and Norbert Wiener*: 268.

34. The situation of one warhead per land-based missile of the strategic forces of the United States and the Soviet Union was inherently stable as a consequence of two important factors. First, "hardening" of the silos in which the missiles were stored yielded the theoretical prediction that under the most favorable conditions imaginable at least two warheads would have to be expended on the average to destroy a single missile in its silo. Second, the United States and the Soviet Union had (and still have) roughly equal numbers of land-based missiles in silos. Thus, even in the most optimistic scenario, to destroy preemptively all of the missiles in the opponent's silos in a surprise attack would require twice as many launchers (missiles) as one had; equivalently, all of one's land-based missiles would be expended destroying just half of the opponent's forces, leaving the other half available for retaliation and nothing in reserve for defense or deterrence. In this situation, the land-

based missiles on both sides represented second-strike retaliatory forces only, thereby contributing to strategic stability. MIRVed missiles, however, can be used as first-strike weapons and so are inherently destabilizing. It is the condition of stability characteristic of one-missile/one-warhead that motivates the argument for the new US Midgetman missile. Of course, the logic is incomplete unless all extant land-based missiles are removed as the single warhead missile is deployed as a replacement.

35. See William C. Selover, "Foot-dragging at SALT?" *The Christian Science Monitor* (May 12, 1970): 1.

36. The official name of the Commission was "The President's Commission on Strategic Forces."

37. See, e.g., Matthew Bunn and Kosta Tsipis, "The Uncertainties of a Preemptive Nuclear Attack," *Scientific American* (November 1983): 38-47.

38. An example of the self-serving influence of government weapons laboratories, in this case with respect to ratification of the Comprehensive Test Ban Treaty (CTB), is reported in R. Jeffrey Smith, "Weapons Labs Influence Test Ban Debate," *Science* 229 (September 13, 1985): 1067-69.

39. Quoted in R. Jeffrey Smith, "Missile Deployments Roil Europe," *Science* 223 (January 27, 1984): 371-76.

40. See the official transcript of *National Security Issues 1981 Symposium* (Bedford, MA: MITRE Corporation, October 13-14, 1981): Document M82-30.

41. Stated Pipes, according to my verbatim transcription (subject to cosmetic editing and added emphasis): "[D]eterrence . . ., to be effective, has to be credible. It will not do to speak of unacceptable damage, a term I've never found a proper definition of – to mean the kind of damage that would be unacceptable to us. You have to develop unacceptable damage in the framework of Soviet thinking. *And that is, above all, not human casualties, which the Russians can bear in very large numbers, nor destructions of property, which they've shown they can assimilate, especially if at stake is the triumph of socialism over capitalism, but weapons, communications and control, and the political system as a whole.* That is the vital nerve, and our strategic package as now devised I think is far more credible as a deterrent for that reason because it's adapted to that way of thinking, than what we had before . . .."

42. See, e.g., Alastair Rae, *Quantum Physics: Illusion of Reality?* (Cambridge, England: Cambridge University Press, 1986): 59.

43. Author's verbatim transcription (subject to cosmetic editing and added emphasis). In New York City in 1984, following a presentation by General Scowcroft on "The Nuclear Arms Race and Public Policy" at the annual meeting of the American Association for the Advancement of Science, I reminded him of the "dilemma" to which he had referred three years earlier and asked what progress, if any, had been made in the interim toward its resolution. He replied: "I think we are gradually moving toward resolving the dilemma, but it is a serious dilemma."

44. Author's verbatim transcription (subject to cosmetic editing and added emphasis).

45. *Ibid.*
46. *Ibid.*
47. The National Command Authorities (NCA) are defined to be the President, the Secretary of Defense, and their deputized alternates and successors. *See* the official transcript of the *National Security Issues 1981 Symposium*: 188.
48. In the United States, the luxury of the imperial presidency must be severely curtailed; the danger is simply too great and the stakes too high to tolerate such free-wheeling independence any longer.
49. As a rule, defense salaries for comparable positions are about 15 percent higher than in the civilian sector. For a more thoroughgoing comparison of defense and civilian employment of scientists and engineers, see Warren F. Davis, "The Pentagon and the Scientist," in John Tirman (ed.), *The Militarization of High Technology* (Cambridge, MA: Ballinger, 1984): 153-79.
50. *See* Lloyd J. Dumas, *The Overburdened Economy* (Berkeley, CA: University of California Press, 1986): 99.
51. Seymour Melman, *The Demilitarized Society: Disarmament and Conversion* (Montreal: Harvest House, 1988): 93.

3

Law and Alternative Security: Toward a Just World Peace

Burns H. Weston

We do not hold the vision of a world without conflict. We do hold the vision of a world without war—and this inevitably requires an alternative system for coping with conflict.

—Adlai E. Stevenson, 1961

With the possible exception of ozone depletion, global warming, and related environmental concerns, nothing is more menacing to the long-term well-being of our planet than the sincerely communicated threat to use nuclear weapons if and when sufficiently provoked. However much nuclear weapons may be rightfully regarded, in their threat role at least, as effective guardians of national security, they portend the severest of consequences to life as we know it. To survive by threatening major extinction is, after all, what we mean by nuclear deterrence.

However, with due respect to the cultural anthropologists who properly counsel against the "killer ape" theory of human nature,[1] conflict is endemic to the human condition,[2] and the average person is not easily dissuaded from a nuclear deterrence system that *seems* to have worked for better than four decades now. People know that the world is and will continue to be filled with conflict, and they want to be protected against its violent excesses. Thus, to escape the mind-boggling risks of nuclear deterrence, thinking about how to ensure world security without relying upon nuclear weapons either extensively or at all is as much a political as it is a moral imperative – in truth, a matter of physical survival. Without an effective alternative to nuclear deterrence, there is no letting go of the nuclear habit, and without letting go of the nuclear habit the world never will be free of the possibility of radioactive annihilation.

Of course, because humankind has the knowledge of how to build nuclear weapons, a knowledge that never can be reversed, it is highly doubtful that the world ever can be made *completely* free of the threat of nuclear war. That is in part why, ever since 1958-59 when the United States Navy argued that the United States required only 232 Polaris missiles to destroy the Soviet Union, numerous concerned observers have advocated minimum deterrence as a means of ensuring national security.[3] Still, given an effective alternative to all forms of nuclear deterrence, with appropriate political will to match, it is possible for the world to be made free of such a threat *almost completely* – to a degree sufficient, at any rate, to eliminate or reduce drastically the current predisposition to rely upon nuclear weapons as a matter of routine, with few safeguards but the willingness of the nuclear weapons States to perceive the common interest of continued human survival inclusively.

Now when contemplating the role of law and legal institutions in this particular regard, it is important to acknowledge, right at the outset, that there really is no such thing as a legal alternative or even a set of legal alternatives to nuclear deterrence. Manifestly, nuclear weapons are weapons of military decisiveness, and as a consequence any substitute for them must be more or less decisive also. Yet, barring some truly radical change in the present State-centric structure of international relations, improbable in the near term (at least outside Western Europe), it seems futile to imagine any legal initiative or set of legal initiatives that *alone* could compete with nuclear weapons as a means of safeguarding core national interests, real or imagined. Embarked though the world surely is upon an historical transformation of major proportion (at least as great as what took place at the end of the Thirty Years' War and the onset of the Westphalian system in 1648), we still live in the *global* Middle Ages, characterized by more than 150 separate fiefdoms, each with a monopoly control over the military instrument and each only barely accountable in any formal sense to each other or to the larger arenas in which each operates. And under such historical circumstances, clearly, it is difficult to secure legal initiatives that *alone* can rise to the challenge of a world essentially free of nuclear weapons.

Which is not to say that international law is without any utility at the present time. As the late Professor William W. Bishop, Jr., counseled pithily over two decades ago, "under present conditions all [States] need international law in order to continue to exist together on this planet."[4] And what is more, it performs, all things considered, remarkably well. Every hour of every day ships ply the sea, planes pierce the clouds, and artificial satellites roam outer space. Every hour of every day communications are transmitted, goods and services traded, and people and things transported from one country to another. Every hour of every day, transactions are made, resources exploited, and institutions created across national and equivalent frontiers. And in all these respects, international law (by which I mean the many processes of authoritative and controlling transboundary decision at all levels of social organization that help to regulate such endeavors) is rather well observed on the whole; it is an important and relevant force in the ordering of human relationships worldwide. True, the international legal system is by no means adequate in its force and effect, and this is particularly true in the realm of war and peace. For example, however much international law may interdict any planned use of nuclear weapons in theory,[5] it scarcely can be said that it is very effective in doing so in practice. But no legal system, not even the most advanced, can boast absolute effectiveness; and all legal systems, again including the most advanced, typically display a certain impotence when it comes to politically volatile or otherwise intractable issues of public policy.[6]

Nor is an admission of the limitations of international law as a substitute for nuclear weaponry the same as saying that international law, fragile though it is at this historical time, cannot or should not play an important role in some alternative strategy to nuclear deterrence. To the contrary, it can and must. But only as part of a larger complex or medley of interdependent policy options – military, technological, economic, political, psychological, and spiritual, as well as legal – that *jointly* can define an integrated, comprehensive, and, above all, decisive security system more or less free of nuclear weapons and the threat of nuclear war.[7] It is not that legal initiatives cannot contribute to an alternative to nuclear deterrence capable of effectively safeguarding core national interests, but that they cannot do so alone.[8] Indeed, no alternative to nuclear deterrence can be other than a composite or medley of interpenetrating policy options if it is to do the job and do it effectively. A post-nuclear international security system must be conceived as an integrated plan.

In the remainder of this chapter, I ask what some of the integral legal initiatives might be. I ask: What legal initiatives might contribute to a global security system more or less free of nuclear weapons and do so effectively?

One obvious answer to this question is of course a genuine and wholesale commitment to the kinds of "deep cuts" and related arms reduction proposals that were contemplated in, for example, the Final Act of the First UN General Assembly Special Session on Disarmament, adopted by consensus in June 1978,[9] and the so-called Five Continent Initiative's Delhi Declaration issued by the Heads of State of Argentina, India, Mexico, Tanzania, Sweden, and Greece in January 1985[10]—proposals that have been the subject of some hopeful but still hesitant discussion between the United States and the Soviet Union at the Strategic Arms Reduction Talks (START) in recent years. Indeed, it would be no small contribution were the superpowers and other actual and would-be nuclear weapons States seriously to commit themselves to the twenty or so principal arms control agreements reached since 1959 and still in force or immediately relevant to the quest for peace in the 1990s— and notwithstanding that, as of this writing, only one, the Intermediate Nuclear Forces (INF) Treaty, reflects any actual dismantling of existing weapons or weapons systems and that many of the agreements prohibit weapons in environments where the military do not particularly want them in the first place.[11]

Another answer, perhaps not so obvious, but one that merits special emphasis given the contempt demonstrated toward the international law of peace during the 1970s and 1980s, not least by the nuclear superpowers,[12] is greatly increased respect for and adherence to the already existing norms of restraint upon the use of force in international relations, as evidenced in, among others, the following instruments:

- the 1928 Kellogg-Briand Pact (or Pact of Paris),[13] UN Charter Article 2(4),[14] the 1965 Declaration on the Inadmissability of Intervention in the Domestic Affairs of States and the Protection of Their Independence and Sovereignty,[15] and the 1970 UN General Assembly Declaration on Principles of International Law Concerning Friendly Relations and Co-operation Among States in accordance with the Charter of the United Nations[16]—all outlawing the non-defensive threat and use of force in the conduct of foreign relations;[17]

- the principles established at Nuremberg and reaffirmed by the United Nations several times since;[18]
- the 1948 Convention on the Prevention and Punishment of the Crime of Genocide;[19] and
- the humanitarian rules of international armed conflict, as embodied in, *inter alia*, the 1907 Hague Regulations Respecting the Laws and Customs of War;[20] the 1925 Protocol for the Prohibition of the Use in War of Asphyxiating, Poisonous or Other Gases;[21] the Hague Draft Rules of Aerial Warfare;[22] the 1949 Convention [No. I] for the Amelioration of the Condition of the Wounded and Sick in Armed Forces in the Field;[23] the 1949 Convention [No. II] for the Amelioration of the Condition of Wounded, Sick and Shipwrecked Members of Armed Forces at Sea;[24] the 1949 Convention [No. III] Relative to the Treatment of Prisoners of War;[25] the 1949 Convention Relative to the Protection of Civilian Persons in Time of War;[26] and Geneva Protocol I Additional Relating to the Protection of Victims of International Armed Conflicts.[27]

Failure or refusal to adhere to these authoritative norms may sometimes yield viscerally pleasing results over the short run, but such disregard rarely if ever yields genuine security, to say nothing of justice, over the long run. In the end, in a world that can easily and quickly "reinvent" nuclear weapons, it is respect — or lack of respect — for the legal prohibitions against the threat and use of force, both defensive and non-defensive, that will determine the fate of the earth.

But what else can be done beyond a commitment to the principal arms control and arms reduction agreements reached since 1959 and to the international law of peace in general? What other legal initiatives might contribute effectively to a global security system more or less free of nuclear weapons?

An answer to this question comes neither easily nor swiftly. One is tempted to take the course of least inventive resistance and recommend the strengthening, where needed, of already existing arms control arrangements — for example, the 1967 Treaty of Tlatelolco establishing a nuclear weapons-free zone in Latin America[28] and the 1968 Nuclear Non-Proliferation Treaty (NPT) prohibiting the direct and indirect transfer of nuclear weapons by nuclear weapons States to non-nuclear weapons States.[29] Each of these non-proliferation regimes lacks, among other things, the full participation of key States critical to their immediate and long-term effectiveness (Argentina, Brazil, Chile, and Cuba in the case of the Treaty of Tlatelolco;[30] China and France in the case of the NPT[31]). Comparable defects attend many of the other existing arms control regimes. Consequently, they too invite recommendations for improvement to enhance global security.

Nevertheless, resisting this relatively easier course, in part because the existing arms control regimes typically presuppose the continued presence of nuclear weapons as a dominant feature of international life, I recommend instead the following series of alternative initiatives as the most worthy of responsible attention as we anticipate the future (drawing heavily upon my colleagues from the Independent Commission on World Security Alternatives,[32] the Lawyers' Committee on Nuclear Policy,[33] and the World Policy Institute[34]). Cast in terms of the normative, procedural, and institutional dimensions of legal process, all presuppose at least three conclusions deducible from an in-depth analysis of the present three- to four-centuries-old international security system:

1. that the present militarily competitive international order cannot be expected to prevent large-scale conventional or nuclear war for very long;
2. that the international system (legal and otherwise) *can* change; and
3. that the prospects for peace and security increase as societies demilitarize, depolarize, denationalize, and transnationalize the global political system.[35]

Explains Robert Johansen, from whom this three-part presupposition is borrowed, the prospects for peace and security increase as security policy "expands the use of non-military power while constricting the use of military power" (demilitarizes); as it "aims to fulfill all human needs in the short- and long-range future" (depolarizes); as it "educates the public to feel a sense of species solidarity" (denationalizes); and as it discourages "state centricity" through public "[p]articipation in intergovernmental and transnational organizations" and private individual and group efforts "to make peace through transnational linkages," preferably "without channeling contacts always through their national capital[s]" (transnationalizes).[36]

NORMATIVE POLICY OPTIONS

Five normative regimes come immediately to mind as capable of helping at least to demilitarize and depolarize the international system.

A Comprehensive Nuclear Weapons Ban

The enforcement of relevant existing international law norms, which interdict virtually any currently planned strategic and tactical use of nuclear weapons,[37] is seriously encumbered by a tradition of political leadership — Machiavellian in character — that typically indulges self-serving interpretations of the legal status of controversial uses of force. A pervasive subjectivity in world politics makes it exceedingly hazardous to tie restraint *vis-à-vis* nuclear weapons to characterizations of warfare as "defensive" or "aggressive," these labels commonly masking politically congenial and politically hostile uses of force. Thus, while perhaps allowing for the possession and retention of some few nuclear weapons for use in the most extreme circumstances, a comprehensive anti-nuclear-weapons regime is needed.[38]

Such a regime would embrace at least the following:

1. an absolute prohibition on first strike and other destabilizing weapons and weapons systems, whether land-based, sea-based, or air-launched — in particular the MX missile, the Trident II (D-5) missile (under development), and the Pershing II and Soviet SS-20 missiles (because such weapons increase the pressure to launch on warning and thereby increase the possibility of nuclear war by accident or miscalculation);[39]
2. a declaration that all research and development (R&D), war plans, strategic doctrines, and strategic threats having first-strike characteristics are illegal *per se*, and that all persons knowingly associated with them are deemed engaged in a continuing criminal enterprise;
3. a presumption that virtually any actual use of nuclear weapons, particularly a first use of such weapons (even in a defensive mode), but also a second or retaliatory "countervalue" use (against cities and other civilian targets), violates the international law of war and constitutes a "crime against humanity";

4.   a clear obligation on the part of all States to pursue nuclear disarmament and otherwise minimize the role of nuclear weapons in inter-State conflict (per Article VI of the NPT[40]) by way of, *inter alia*,
  (a)   a renunciation of the *policy* of first use and the war-fighting doctrines and capabilities that accompany it,
  (b)   a comprehensive nuclear test ban (CTB),[41] and
  (c)   strengthened nuclear non-proliferation regimes;[42]
5.   a commitment to a strengthened Anti-Ballistic Missile Treaty[43] (because pursuit of an anti-ballistic missile defense system stimulates competition in offensive weapons) together with a ban on all space weapons and space-based missile defense systems (because such systems, especially if not preceded by deep cuts in offensive ballistic missiles, are likely to encourage a proliferation of the most destabilizing weapons and weapons systems); and
6.   a clear mandate for all citizens to take whatever steps may be available to them, including acts of nonviolent civil disobedience, to expose the illegality of the use of nuclear weapons and to otherwise insist upon the lawful conduct of the foreign policies of their own governments.[44]

A comprehensive nuclear weapons ban such as this, it should be understood, genuflects ever so slightly in the direction of a policy of minimum deterrence[45] because it allows for the use of some nuclear weapons in extreme circumstances. However, the uses contemplated here are only those that possibly, but not altogether unambiguously, may be deemed permissible under international law, *i.e.*, (a) very limited tactical—mainly battlefield—warfare utilizing low-yield, "clean," and reasonably accurate nuclear weapons for second-use, retaliatory purposes only; and (b) an extremely limited "counter-force" strike in strategic and theater-level settings for second-use retaliatory purposes only (until as yet undeveloped technological refinements are achieved).[46] The minimum deterrence strategies currently popularly advocated appear not to fit within these parameters and consequently are morally as well as legally suspect.

A Comprehensive Ban on *Non-Nuclear* Weapons of Mass Destruction

The same arguments that warrant a comprehensive ban on nuclear weapons compel also a comprehensive ban on non-nuclear weapons of mass destruction, *including chemical and biological weapons*. In addition, a ban on such weapons would lessen the prospect that a belligerent State, especially a beleaguered one, might establish or renew a dependence upon nuclear weapons.

Such a ban would include at least the following:

1. an absolute prohibition on the development, production, stockpiling, and use of conventional mass destruction weapons and weapons systems, including chemical and biological weapons of mass destruction;
    (a) a strengthened Geneva Gas Protocol,[47] prohibiting the possession as well as the use of the gas and bacteriological methods of warfare covered by the Protocol;
    (b) a strengthened Biological Weapons Convention,[48] providing for effective on-site inspections and enforcement mechanisms capable of responding to scientific advances and new biological technologies to ensure against violations of the Convention;
2. a declaration that all R&D, war plans, strategic doctrines, and strategic threats having non-nuclear mass destructive characteristics are illegal *per se*, and that all persons knowingly associated with them are deemed engaged in a continuing criminal enterprise;
3. a presumption that virtually any actual use of non-nuclear weapons of mass destruction, particularly a first use of such weapons (even in a defensive mode), but also a second or retaliatory use (against cities and other civilian targets), violates the international law of war and constitutes a "crime against humanity";
4. a clear obligation on the part of all States to eliminate all non-nuclear weapons of mass destruction from their arsenals, including chemical and biological weapons of mass destruction, and otherwise to minimize the role of such weapons in inter-State conflict; and
5. a clear mandate for all citizens to take whatever steps may be available to them, including acts of nonviolent civil disobedience, to expose the illegality of the use of non-nuclear weapons of mass destruction and to otherwise insist upon the lawful conduct of the foreign policies of their own governments.

The point of these limitations, it should be understood, is to restrict all military strategy to a non-offensive/non-provocative defense posture exclusively.[49] Only such an arrangement, it is believed, will permit States to resist the temptation to resort to nuclear weapons.

A Conventional Weapons Non-Proliferation Regime

Just as there has been a proliferation of nuclear weapons since 1945, so also has there been a proliferation in the manufacture and export of conventional weapons, particularly to the Third World.[50] This fact is well known.[51] Yet, notwithstanding that this traffic in conventional arms increases not only the destructiveness of conflict but also the likelihood of bloody conflict erupting, the world community stands by and does essentially nothing.

The world community does so, however, at great peril to itself— indeed, far greater peril than commonly is realized. Just as conventional arms are "trip wires" to conventional wars, so are conventional wars—and their arms—"trip wires" to nuclear conflict, capable of engaging the superpowers and other major powers and thereby risking escalation to nuclear war. In the absence of a ban on the manufacture and export of conventional weapons, a post-nuclear world would be similarly endangered. To the extent that, in such a world, conventional wars could seriously jeopardize the real and perceived interests of nuclear-prone States, so too could they serve as catalysts to the "reinvention" and subsequent actual use of nuclear weapons to safeguard those interests.

Thus, a conventional weapons non-proliferation regime, greatly limiting if not altogether prohibiting conventional arms traffic, would seem as much a necessity to a post-nuclear global security system as the existing nuclear non-proliferation regime is to the present-day nuclear deterrence system. It seems particularly a necessity relative to such large and actually or potentially provocative weapons and weapons systems as tanks, armored cars, warships, long-range "attack" aircraft, missiles, and other components of "forward defense." These large weapons and weapons systems, in addition to being the most easily regulated because they are the most easily detected, are the most capable of conventional weapons systems of contributing to mass destruction. At the very least, such a regime should ensure an effective surveillance and record-keeping system, capable at least of alerting responsible elites of the presence of dangerous world practices and trends.

A NATO-Warsaw Pact Non-Aggression Regime

The question of across-the-board reductions in strategic and other nuclear forces is commonly seen as inextricably linked to a claimed comparative advantage in conventional forces on the part of the Warsaw Pact countries relative to the NATO powers. Accordingly, a NATO-Warsaw Pact non-aggression regime would seem a minimal necessity for a nuclear weapons ban and, indeed, for a post-nuclear security system as a whole. As former US Ambassador to the Soviet Union and "cold warrior" — now elder statesman — George F. Kennan "confessed" a few years ago:

> I am now bound to say that while the earliest possible elimination of Soviet weaponry is of no less importance in my eyes than it ever was, this would not be enough, in itself, to give Western civilization even an adequate chance of survival. War itself, as a means of settling differences at least between the great industrial powers, will have to be in some way ruled out. . . .[52]

In other words (recalling a point made earlier), to rid ourselves of the *nuclear* habit we must rid ourselves also of the *war* habit.

Kennan's "confession" is, of course, a prescription for basic system transformation worldwide, an ambition understandably daunting to the average mind. A logical place to begin, however, would be in the negotiation and conclusion of a mutual non-aggression regime between the NATO and WTO (Warsaw Pact) countries that, in turn, would facilitate — and optimally be accompanied by — the dismantling of the vast military establishments of the major powers (in particular, the US, the USSR, and the two Germanys) that heretofore have provided, in substantial part, at least the formal excuse for the invention and deployment of nuclear weapons and their supporting systems.[53] Also, it could serve as a model for agreements between other proven or potential antagonists (*e.g.*, Iran-Iraq). Given the favorable disposition toward nuclear weapons and other weapons of mass destruction exhibited by some of the world's lesser powers, the negotiation and conclusion of mutual non-aggression regimes between other antagonists would be likewise desirable and helpful.

Integral to a NATO-WTO (or other) non-aggression regime and a logical end point for the mutual force reduction (MFR) talks begun in Vienna in 1973 and recently rekindled in the NATO-WTO Negotiations on Conventional Armed Forces in Europe (CAFE or CFE) would be a call for the establishment of non-offensive/non-provocative defense ceilings on troop levels and accompanying force structures authorized to each side[54]—which is to say, both quantitative and qualitative disarmament as between East and West and especially as between the two nuclear superpowers and the two Germanys. The goal would be not merely numerical reductions in troop levels and weapons (quantitative conventional disarmament) but increased stability via a reduction in the risk of surprise attack or uncontrolled escalation lest somehow war should break out between the two sides (qualitative conventional disarmament).[55] Central to a NATO-WTO (or other) non-aggression regime would be a shift in each side's force structure from offensive to mutual defensive postures (akin to what has been adopted in modified form in Switzerland, Sweden, and Yugoslavia), which would greatly reduce the risk of war, including escalation to a nuclear war, because of the "mutual defensive superiority" of each side.

Also integral to a NATO-WTO (or other) non-aggression regime would be a call for such "confidence-building measures" (CBMs) as: (1) a required notification of all "mass" take-offs of bombers from each side; (2) a ban on all close military approaches to either side's territory or territorial waters in air and naval exercises; (3) a ban on all ground maneuvers adjacent to any East-West border involving more than, say, 25,000 military personnel; (4) a withdrawal of all "forward defense" weapons from a 150- to 300-mile corridor along all East-West (or other) borders; and (5) a requirement of consistent exchange of military data, verifiable by on-site inspection, relative to numbers of active and reserve troops, quantities and locations of conventional (especially dual-capable) weapons, deployment of force structures, and so forth. The negotiations on Confidence- and Security-Building Measures (CSBM) under way in Vienna as of this writing would be a logical forum within which to consider such recommendations, where indeed some of them already are being considered.

A US-USSR Non-Intervention Regime

It is clear, to the informed citizen at least, that the current nuclear deterrence system operating between the two superpowers is in reality a system of extended deterrence, meant to guard against far more than European battlefield confrontations and intercontinental strategic attacks. Both the United States and the Soviet Union have strong hegemonic interests in keeping the other out of their respective spheres of influence and, beyond that, out of so-called non-aligned countries. But it is clear, too, that this extended or hegemonic deterrence system is of necessity nuclear because the economies of neither the United States nor the Soviet Union can afford, without major domestic sacrifice, a conventional one, a fact seemingly well understood at least by President Gorbachev in recent years. Thus, because the strong economic and political interests of the two superpowers simply will not go away — and, indeed, may become even more "vital" to them, at least psychologically speaking, as they ever more discover that they are unable to control people and events as they once did — it is essential that a post-nuclear global security system make as one of its cornerstones a compact by each government to refrain from sending any of its armed forces into the other's clear sphere of influence (however defined) or into any non-aligned country, even if invited. At the risk of implying a recognition of the other's "right" to rule in its sphere of influence, a mutual promise of self-restraint on the part of both superpowers, especially one that would ensure their observance of the territorial integrity and political independence of Third World countries, would go a long way toward guaranteeing the viability of a post-nuclear global security system. Where force may be needed to prevent or minimize deprivations of fundamental human rights and freedoms,[56] then there should be recourse to the global and regional intergovernmental organizations that are designed to police such matters on a multilateral basis.[57]

PROCEDURAL POLICY OPTIONS

A global security system that forswears reliance upon nuclear weapons can provide no security at all without clearly established and respected procedures for both peacekeeping and peacemaking. If inter-State disputes can be prevented from degenerating into armed hostilities or settled by peaceful means, they are unlikely to escalate into threats to the peace or acts of aggression and war. It is true that the past record of undertakings to keep the peace under the aegis of the United Nations and to achieve dispute settlement through international tribunals, arbitration, and similar peaceful means has not been very encouraging. But established and respected procedures for multilateral peacekeeping and for the mediation, conciliation, arbitration, and adjudication of international disputes, preferably within the framework of the United Nations but desirably also at the regional level, would seem nevertheless necessary even if not sufficient for the maintenance of world peace and security.

Without the active participation of States in peaceful efforts to accommodate each other, there is little likelihood of achieving the stability and harmony that a world free of nuclear weapons would require.

Thus, the following modest procedural initiatives would seem necessary and useful (perhaps especially at the early stages of international accommodation and nuclear disarmament).

Improvement of UN Peacekeeping Opportunities and Capabilities

UN peacekeeping opportunities and capabilities can be improved by:[58]

1. *facilitating automatic peacekeeping action* on the basis of predetermined levels of crisis or thresholds of conflict (possibly variable, depending on the geopolitical context), thus avoiding the obstructions posed by the exercise of the Security Council veto;

2. *assuring peacekeeping finances* on an automatic basis, possibly through a percentage surcharge added to annual assessed contributions, thus again avoiding the obstructions posed by the exercise of the Security Council veto;

3. *guaranteeing military units* (land, sea, and air) on a more or less permanent standby basis (as envisaged in UN Charter Article 43), trained for peacekeeping by the member States in the course of their militaries' basic training and on the basis of expertise and additional training provided by an appropriate UN agency;

4. *regularly stockpiling military equipment and supplies* needed to enhance the UN's capacity to undertake peacekeeping operations on short notice;

5. *ensuring access to conflict areas* without requiring the initial or continuing permission of the conflicting parties; and

6. *tying UN peacekeeping to peacemaking* (*i.e.*, pacific settlement) to ensure that the merits of any given dispute will receive the attention that is needed to achieve long-term stability in the troubled area.

Improvement of UN and Other Peacemaking Opportunities and Capabilities

UN peacemaking opportunities and capabilities can be improved by:[59]

1. *enhancing and making greater use of UN dispute settlement mechanisms*, most of which have been rarely if ever used (such as the General Assembly's 1949 Panel for Inquiry and Consultation, a 1950 Peace Observation Commission, and a 1967 Register of Experts on Fact-Finding);
2. *encouraging increased consent to mediation, conciliation, arbitration, and adjudication* via
    (a) guarantees limiting the scope of the third-party judgment to the determination of the doctrines, principles, and rules that could guide the parties in approaching settlement; and
    (b) greater use of technically non-binding advisory opinions;
3. *increasing reliance on private persons and non-governmental organizations (NGOs) as neutral intermediaries* (thereby helping to avoid escalating arguments to full-scale inter-State disputes)
    (a) in pre-dispute consultations and in post-dispute negotiated settlements; and
    (b) before international tribunals for the purpose of clarifying a customary law norm or a clause in an international agreement; and
4. *convening periodic regional conferences on security and cooperation* similar to the one launched in Helsinki for Europe in 1975 — such conferences to reflect the priorities and circumstances of the separate regions and, with help from the UN Secretariat, to serve the essential decision function of appraisal and recommendation not only on matters relating directly to international security but on economic, social, and cultural matters upon which international security commonly depends; and
5. *adopting a code of international peacemaking procedures* (drawn from a variety of existing instruments) that would allow governmental officials to develop confidence in available procedures and that States could accept as binding upon them in whole or in part.

Improvement of Opportunities and Capabilities for Legal Challenges to Coercive Foreign Policies

Opportunities and capabilities for legal challenges to coercive foreign policies can be improved by:

1. *enhancing the role of the International Court of Justice*[60] relative to threats to the peace, breaches of the peace, and acts of aggression through, for example,
   (a) expanded acceptance of the Court's compulsory jurisdiction and greater use of its advisory jurisdiction relative to actual or potential hostilities between States;
   (b) broadened standing to petition the Court to permit access by qualified non-governmental organizations;
   (c) increased appeal to the Court's specialized "chamber procedure" in respect of inter-State conflicts unresolved by more local remedies;[61]
2. *facilitating application of the international law of peace in domestic courts* through, for example,
   (a) the reduction of barriers to "legal standing" on the part of private litigants especially; and
   (b) the narrowing of doctrines of non-justiciability (*e.g.,* the "political question," "act of State," and "sovereign immunity" doctrines) to encourage public accountability in the conduct of foreign policy.

Of course, all of these and similar procedural initiatives have their share of difficulties: winning the confidence of contentious sovereign powers; achieving genuine neutrality in disputes; maintaining effective communication; overcoming legal and political isolationism; and so forth. Nonetheless, all are worthwhile because they would directly help to demilitarize, depolarize, denationalize, and transnationalize the international system—outcomes that, as indicated above, enhance the prospects for international peace and security.[62]

INSTITUTIONAL POLICY OPTIONS

At least six institutional responses recommend themselves to a post-nuclear security system, each directly helping—some perhaps within the framework of the United Nations, some perhaps not—to depolarize, denationalize, and transnationalize the international system and thereby to contribute to a more peaceful world order.[63] They of course do not exhaust the institutional policy options that might be recommended.

1. *Establish an international disarmament verification agency* which, through satellite observation, seismic and atmospheric surveillance, and on-site inspection, could supplement national means of verification and be capable of transnational monitoring of world military capabilities and movements.[64] Such an agency, with a membership comprising non-nuclear as well as nuclear weapons States, would (a) oversee the implementation of arms control and arms reduction agreements; (b) provide an impartial means of detecting and guarding against the secret testing and production of nuclear weapons and other weapons of mass destruction, including chemical and biological weapons; (c) discourage provocative military buildups and maneuvers; and (d) otherwise acquire the vital experience and reliability needed if arms reductions are ever to proceed very far. As a means of achieving genuine effectiveness, it also would be expected to establish regional oversight boards with authority to conduct on-site inspections of any and all weapons-capable facilities at the request of any State party or qualified non-governmental organization.

2. *Create an international technological development and weapons program agency* to (a) foster joint research of defensive technologies by multilateral teams of scientists and (b) prevent and restrain arms buildups. Such an organ would, for example, provide the Soviet Union and the United States with the opportunity to share defensive technology and to facilitate missile defense research without imperiling the ABM Treaty[65] or otherwise exacerbating the arms race. Also, it could reduce inclinations to surprise the other side with new and threatening developments. As such it could help solidify the new turning point in world affairs that has come about with the advent of *glasnost* and *perestroika*.

3. *Create risk-reduction opportunities and capabilities* by establishing, for example: (a) a joint inter-State consultation commission with a permanent staff composed of the nationals of disputing parties (among others) capable of handling actual and potential conflicts by way of routine review rather than the usual procedure of consulting only in extraordinary circumstances; (b) a joint inter-State negotiating commission composed of nationals from each side of a conflict, working together to find a solution acceptable to all concerned; (c) regional mediation, conciliation, and arbitration

panels composed of persons of recognized competence and fair-mindedness with authority to investigate and seek the resolution of conflicts and disputes otherwise capable of culminating in hostilities. Where these "local remedies" do not succeed, then appeal should be had to the International Court of Justice for final and binding resolution of the disputes in question.[66] In any event, the common primary purpose of these risk-reduction remedies would be to facilitate communication between contending parties to avert the possibility of war through miscalculation or misperception.

4. *Create an international "weapons into plowshares" agency* through which the conversion of national arms industries to socially redemptive production could be facilitated and a concrete connection between those who spend resources on armaments and those in economic and technological need could be fruitfully established.[67] The overriding purpose of such an agency, which among other things could help bring labor unions and industrial management together in common enterprise, would be to encourage a comprehensive process of reconstruction and renewal conducive to the establishment of a genuinely productive and equitable world economy that, in turn, would greatly reduce the likelihood that nations would do military battle with one another.[68]

5. *Create permanent global or regional police forces*[69] consisting of persons recruited individually instead of from national military contingents (as in past UN peacekeeping experience), each with loyalty to world or regional authorities rather than national authorities. Such forces would be unencumbered by divided loyalties and by the possibility of sudden, unanticipated recall or withdrawal by national governments (as has happened with *ad hoc* UN forces in the Middle East, for example). As a consequence, they would be more readily available, more subject to efficient coordination, and thus more effective overall. As such, better positioned to establish useful precedents over time, they would constitute a further significant step in assuring a successful security system not dependent on nuclear weapons. Of course, appropriate precautions would have to be taken to guard the guardians.

6. *Create a permanent international criminal court* with compulsory jurisdiction specifically over war crimes, crimes against the peace, and crimes against humanity, accessible by multilateral intergovernmental organizations, non-governmental entities, and qualified individuals, as well as by States. The need for an international judicial body to try violations of international criminal

law, either as a chamber of the International Court of Justice or as an independent entity, has been recognized for years.[70]

In addition to these six institutional initiatives one might of course mention the possibility that the United Nations, because it mirrors the power structure that existed at its birth, may have outlived its usefulness as a security system and therefore should be replaced by a new international authority (or cluster of authorities) that reflect(s) the growing aspiration for a restructured international order that possesses strengthened conflict resolution capabilities. However, so sweeping an initiative seems beyond the capacity of existing international actors and, in any event, dismisses too quickly the more immediate possibility of UN reform and collective security under Security Council or other auspices.[71] Moreover, it is a precarious option for the foreseeable future inasmuch as it could be deviously used, out of some nostalgic yearning for a bygone era, to banish the United Nations without in any way providing for its replacement.   Still, the possibility of a new world authority should not be overlooked.[72]

CONCLUSION

Thus it is evident that there are a number of possible *legal* initiatives that might contribute effectively to a nuclear-weapons-free global security system—but only, as indicated, as part of an integrated plan.

It bears emphasis, however, that all of the above recommendations are the logical outgrowth of a lexicology that defines security—personal, national, and international—virtually exclusively in terms of the absence of war or the threat of war; and as a consequence they bespeak the norms, procedures, and institutions that facilitate the prevention or elimination of *military* confrontation and conflict. As became increasingly clear from the worldwide economic and environmental pressures of the 1970s and 1980s, however, a definition of security informed preeminently by concern for military risks and encounters is not adequately responsive to the full range of threats to our personal, national, and international security that already we have encountered and that we are likely to encounter in the 1990s and the years after 2000 as well.[73] Achieving true global security will require not only a drastic circumscription of nuclear and, more generally, militarist tendencies, but also the progressive development of those legal norms, procedures, and institutions that can assist the promotion and protection of social justice, economic well-being, and ecological balance on a worldwide scale. It is social injustice, economic malaise, and environmental decline that lead, independently and interdependently, to frustration, conflict, and oftentimes, ultimately, violence.[74] The evidence is all around us. Therefore, a post-nuclear global security system is unlikely to succeed if it is not marked also by a broad and deep commitment to the widespread realization of fundamental human rights and freedoms, to the wholesale eradication of grinding poverty and economic dependency, and to the unwavering stewardship of our earth-space environment (the oceans, the atmosphere, the biosphere, and the geosphere) as a total living organism, meant to be cherished rather than squandered.[75] And to these ends, of course, including the repeal of the parochial, piecemeal, and timorous policies that have allowed ours to become a seriously endangered planet, there is vast room for law and lawyering, both domestic and international.[76]

On the other hand, it also bears emphasis that it is not, on final analysis, treaties and charters prescribing specific norms, procedures, and institutions that will guarantee an enduring condition of peace among nations. It is, rather, the ingrained assumptions and habits of men and women everywhere, above all men and women in government and other arenas of social responsibility, that ultimately will be determinative in this regard. And if an international security system that consciously abjures reliance upon nuclear weapons is to succeed, then these assumptions and habits will have to move beyond the present, singular focus on *national* security to the wider notion of *global* security, now made mandatory by economic and environmental strains that increasingly are transcending national frontiers and eroding the sacred boundaries of national sovereignty. The entire human race – not one territorial constituent of it – must become the conscious beneficiary of all alternative security initiatives. A sense of species solidarity and a concern for all peoples, not just the ruling elites, must underwrite all proposals for alternative security as we proceed, in the words of Jesuit philosopher Pierre Teilhard de Chardin, in "the planetization of Mankind."[77]

All this said, however, a haunting question hovers over this analysis, just as it hovers over every analysis about alternatives to nuclear deterrence: Has humankind the acumen and the political will to do something constructive before it is too late? That is *the* real issue in these and all related discussions at the present time. And in this connection it is helpful to recall the words of the late Loren Eiseley in his splendid little book *The Immense Journey*, as much a work of poetry as of anthropology. "Within the reasonable limits of the brain that now exists," Eiseley wrote,

> [nature] has placed the long continuity of civilized memory as it lies packed in the world's great libraries. The need is not really for more brains, the need is now for a gentler, a more tolerant people than those who won for us against the ice, the tiger, and the bear.[78]

The haunting question that hovers over us, in other words, is whether we are or can be the gentler, the more tolerant, people of which Eiseley speaks. If so, then a new international security is a serious possibility.

NOTES

An earlier version of this essay was presented at the First World Congress of the International Association of Lawyers Against Nuclear Arms (IALANA) at The Hague, The Netherlands, September 22-24, 1989. For comments thereon, I am indebted to Richard Falk, Rein Müllerson, and Bruce Russett.

1. *See, e.g.,* Alexander Alland, *The Human Imperative* (New York: Columbia University Press, 1972); René Dubos, *Beast or Angel? Choices That Make Us Human* (New York: Scribner, 1974).
2. *See* R. Paul Shaw and Yuwa Wong, *Genetic Seeds of Warfare: Evolution, Nationalism, and Patriotism* (Boston: Unwin Hyman, 1989).
3. Minimum deterrence, although representing a major rollback in humankind's ability to self-destruct, would not eliminate the war-making capabilities of any nation. However, it would mean the elimination of many thousands of nuclear weapons from the world's nuclear arsenals and thus reduce threat and increase stability. *See, e.g.,* Morton Halperin, *Nuclear Fallacy* (Cambridge, MA: Ballinger, 1987); Robert S. McNamara, *Blundering Into Disaster* (New York: Pantheon, 1986); Vincent Ferraro and Kathleen FitzGerald, "The End of a Strategic Era: A Proposal for Minimal Deterrence," *World Policy Journal* 2 (Winter 1984): 339. For summary discussion, see Harry B. Hollins, Averill L. Powers, and Mark Sommer, *The Conquest of War: Alternative Strategies for Global Security* (Boulder, CO: Westview, 1989): 54-63. *See also* this chapter: A Comprehensive Nuclear Weapons Ban.
4. William W. Bishop, Jr., "General Course of Public International Law," *Receuil des Cours* 115 (Hague Academy of International Law, 1965-II): 147, 467.
5. *See,* in particular, Burns H. Weston, "Nuclear Weapons Versus International Law: A Contextual Reassessment," *McGill Law Journal* 28 (1983): 542. *See also* Daniel Arbess, "The International Law of Armed Conflict in Light of Contemporary Deterrence Strategies: Empty Promises or Meaningful

Restraint?" *McGill Law Journal* 30 (1984): 89; Francis A. Boyle, "The Relevance of International Law to the 'Paradox' of Nuclear Deterrence," *Northwestern University Law Review* 80 (1986): 1407; Elliott L. Meyrowitz, "The Opinions of Legal Scholars on the Legal Status of Nuclear Weapons," *Stanford Journal of International Law* 24 (1988): 111.

6. Reflect, for example, on the interrelation of law and the "war on drugs" in the United States or on the real possibilities of Presidential impeachment under the United States Constitution.

7. *See, e.g.,* the other chapters in this volume.

8. As J. L. Brierly wrote, international law "is neither a myth on the one hand, nor a panacea on the other, but just one institution among others which we can use for the building of a better international order." J. L. Brierly, *The Law of Nations* (New York and Oxford: Oxford University Press, Sixth Edition, 1963): v.

9. June 30, 1978, U.N.G.A. Res. S-10/2 (S-X), 10 (Special) U.N. GAOR, Supp. (No. 4) 3, U.N. Doc. A/S-10/4 (1978), *reprinted in* Burns H. Weston, Richard A. Falk, and Anthony D'Amato, *Basic Documents in International Law and World Order* (St. Paul, MN: West Publishing, Second Edition, 1990); hereinafter *Basic Documents*.

10. *See Conference on Disarmament*, U.N. Doc. CD/549 (February 6, 1985), *reprinted in Basic Documents*: 273-74.

11. *See* the Antarctic Treaty, December 1, 1959, 12 U.S.T. 794, T.I.A.S. No. 4780, 402 U.N.T.S. 71; the Memorandum of Understanding Between the United States and the Union of Soviet Socialist Republics Regarding the Establishment of a Direct Communication Link ("the Hot Line Agreement"), June 20, 1963, 14 U.S.T. 825, T.I.A.S. No. 5362, 472 U.N.T.S. 163; the Treaty Banning Nuclear Weapon Tests in the Atmosphere, in Outer Space and Under Water ("the Partial Test Ban Treaty"), August 5, 1963, 14 U.S.T. 1313, T.I.A.S. No. 5433, 480 U.N.T.S. 43, *reprinted in International Legal Materials* 2 (1963): 889; the Treaty on Principles Governing the Activities of States in the Exploration and Use of Outer Space, Including the Moon and Other Celestial Bodies ("the Outer Space Treaty"), January 27, 1967, 18 U.S.T. 2410, T.I.A.S. No. 6347, 610 U.N.T.S. 205, *reprinted in International Legal Materials* 6 (1967): 386; the Treaty for the Prohibition of Nuclear Weapons in Latin America ("the Treaty of Tlatelolco"), February 14, 1967, 634 U.N.T.S. 281, *reprinted in International Legal Materials* 6 (1967): 521; the Treaty on the Non-Proliferation of Nuclear Weapons ("the NPT"), July 1, 1968, 21 U.S.T. 483, T.I.A.S. No. 6839, 729 U.N.T.S. 161, *reprinted in International Legal Materials* 7 (1968): 811; the Treaty on the Prohibition of the Emplacement of Nuclear Weapons and Other Weapons of Mass Destruction on the Seabed and the Ocean Floor and in the Subsoil Thereof ("the Seabed Arms Control Treaty"), February 11, 1971, 23 U.S.T. 701, T.I.A.S. No. 7337, 955 U.N.T.S. 115, *reprinted in International Legal Materials* 10 (1971): 146; the Agreement on Measures to Reduce the Risk of Outbreak of Nuclear War Between the United States and the Union of Soviet Socialist Republics ("the Accident Measures Agreement"), September 30, 1971, 22 U.S.T. 1590, T.I.A.S. No. 7186,

807 U.N.T.S. 57, *reprinted in International Legal Materials* 10 (1971): 1173; the Agreement Between the United States of America and the Union of Soviet Socialist Republics on Measures to Improve the USA-USSR Direct Communications Link ("the Hot Line Modernization Agreement"), September 30, 1971, 22 U.S.T. 1598, T.I.A.S. No. 7187, 806 U.N.T.S. 402, *reprinted in International Legal Materials* 10 (1971): 1174; the Convention on the Prohibition of the Development, Production and Stockpiling of Bacteriological (Biological) and Toxin Weapons and on Their Destruction, April 10, 1972, 26 U.S.T. 583, T.I.A.S. No. 8062, 1015 U.N.T.S. 163, *reprinted in International Legal Materials* 11 (1972): 310; the Treaty Between the United States of America and the Union of Soviet Socialist Republics on the Limitation of Anti-Ballistic Missile Systems ("the ABM Treaty"), May 26, 1972, 23 U.S.T. 3435, T.I.A.S. No. 7503, 944 U.N.T.S. 13, *reprinted in International Legal Materials* 11 (1972): 784; the Interim Agreement Between the United States of America and the Union of Soviet Socialist Republics on Certain Measures with Respect to the Limitation of Strategic Offensive Arms, May 26, 1972, 23 U.S.T. 3462, T.I.A.S. No. 7504, 94 U.N.T.S. 3, *reprinted in International Legal Materials* 11 (1972): 791; the Declaration of Basic Principles of Relations Between the United States of America and the Union of Soviet Socialist Republics, May 29, 1972, *Department of State Bulletin* 66 (1972): 898; the Treaty Between the United States of America and the Union of Soviet Socialist Republics on the Limitation of Underground Nuclear Weapon Tests, July 12, 1974, *Department of State Bulletin* 71 (1974): 217; the Limitations on Anti-Ballistic Missile Systems Treaty Protocol, July 3, 1974, 27 U.S.T. 1645, T.I.A.S. No. 8276; the Joint Statement on the Limitation of Strategic Offensive Arms ("the Vladivostok Agreement"), April 29, 1974, *Department of State Bulletin* 70 (1974): 677; the Final Act of the Conference on Security and Cooperation in Europe ("the Helsinki Accords"), August 1, 1975, *Department of State Publication No. 8826* (General Foreign Policy Series 298), *reprinted in International Legal Materials* 14 (1975): 1292; and the Treaty Between the United States of America and the Union of Soviet Socialist Republics on the Limitation of Strategic Offensive Arms and Protocol Thereto ("the SALT II Treaty), June 18, 1979, S. Exec. Doc. Y, 96th Cong, 1st Sess. 37 (1979); Agreement Governing the Activities of States on the Moon and Other Celestial Bodies ("the Moon Treaty"), December 5, 1979, U.N.G.A. Res. 34/68 (XXXIV), 34 U.N. GAOR, Supp. (No. 46) 77, U.N. Doc. A/34/664 Annexes (1979), *reprinted in International Legal Materials* 18 (1979): 1434; Treaty Between the United States of America and the Union of Soviet Socialist Republics on the Elimination of Their Intermediate-Range and Shorter-Range Missiles ("INF Treaty"), December 8, 1987, *Department of State Publication No. 9555* (December 1987), *reprinted in International Legal Materials* 27 (1988): 90. Most of the foregoing agreements are reprinted in whole or in part in *Basic Documents*.

   12. *See,* for example, Burns H. Weston, "The Reagan Administration Versus International Law," *Case Western Reserve Journal of International Law* 19 (1987): 295.

13. Treaty Providing for the Renunciation of War as an Instrument of National Policy, August 27, 1928, 46 Stat. 2343, T.S. No. 796, 2 Bevans 732, 94 U.N.T.S. 57, *reprinted in Basic Documents*: 137.

14. June 26, 1945, 59 Stat. 1031, T.S. No. 993, 3 Bevans 1153, 1976 Y.B. U.N. 1043, *reprinted in Basic Documents*: 16-32.

15. December 21, 1965, U.N.G.A. Res. 2131 (XX), 20 U.N. GAOR, Supp. (No. 14) 11, U.N. Doc. A/6014 (1966), *reprinted in Basic Documents*: 195-96.

16. October 24, 1970, U.N.G.A. Res. 2625 (XXV), 25 U.N. GAOR, Supp. (No. 28) 121, U.N. Doc. A/8028 (1971), *reprinted in Basic Documents*: 108-13.

17. *See also* Articles 8 and 10-11 of the Convention on the Rights and Duties of States, December 26, 1933, 49 Stat. 3097, T.S. No. 881, 3 Bevans 145, 165 L.N.T.S. 19 (also known as "the Montevideo Convention"); Article 5 of the Pact of the League of Arab States, March 22, 1945, 70 U.N.T.S. 237; Chapters II, IV and V of the Charter of the Organization of American States, April 30, 1948, 2 U.S.T. 2394, T.I.A.S. No. 2361, 119 U.N.T.S. 3 and Protocol of Amendment, February 27, 1967, 21 U.S.T. 607, T.I.A.S. No. 6847; the Inter-American Treaty of Reciprocal Assistance, September 2, 1947, 62 Stat. 1681, T.I.A.S. No. 1838, 121 U.N.T.S. 77 (also known as "the Rio Pact"); the American Treaty on Pacific Settlement, April 30, 1948, 30 U.N.T.S. 55 (also known as "the Pact of Bogota"); and Articles II and III of the Charter of the Organization of African Unity, May 25, 1963, 479 U.N.T.S. 39. All of these instruments are reprinted in whole or in part in *Basic Documents*.

18. *See* Article 6 of the Charter of the International Military Tribunal, August 8, 1945, 59 Stat. 1544, 82 U.N.T.S. 279, *reprinted in Basic Documents*: 138-39; Judgment of the International Military Tribunal, Nuremberg, International Military Tribunal, Trial of the Major War Criminals 1 (1947): 171; Affirmation of the Principles of International Law Recognized by the Charter of the Nuremberg Tribunal, December 11, 1946, U.N.G.A. Res. 95(I), U.N. Doc. A/236 (1946), *reprinted in Basic Documents*: 140; Declaration on the Prohibition of the Use of Nuclear Weapons and Thermo-nuclear Weapons, November 24, 1961, U.N.G.A. Res. 1653 (XVI), 16 U.N. GAOR, Supp. (No. 17) 4, U.N. Doc. A/5100 (1961), *reprinted in Basic Documents*: 190-91.

19. December 9, 1948, ___ U.S.T. ___, T.I.A.S. No. ___, 78 U.N.T.S. 277, *reprinted in Basic Documents*: 297.

20. Annex to the 1907 Hague Convention [No. IV] Respecting the Laws and Customs of War on Land, October 18, 1907, 36 Stat. 2277, T.S. No. 539, 1 Bevans 631, *reprinted in Basic Documents*: 129-35.

21. June 17, 1925, 26 U.S.T. 575, T.I.A.S. 8061, 94 L.N.T.S. 65 (the "Geneva Gas Protocol"), *reprinted in Basic Documents*: 136.

22. *Reprinted in American Journal of International Law* 17 (1923, Supplement): 245.

23. August 12, 1949, 6 U.S.T. 3114, T.I.A.S. No. 3362, 75 U.N.T.S. 31, *reprinted in Basic Documents*: 147-54.

24. August 12, 1949, 6 U.S.T. 3217, T.I.A.S. No. 3363, 75 U.N.T.S. 85, *reprinted in Basic Documents*: 155-59.

25. August 12, 1949, 6 U.S.T. 3316, T.I.A.S. No. 3364, 75 U.N.T.S. 135, *reprinted in Basic Documents*: 160-69.

26. August 12, 1949, 6 U.S.T. 3516, T.I.A.S. No. 3365, 75 U.N.T.S. 287, *reprinted in Basic Documents*: 170-80.

27. December 12, 1977, U.N. Doc. A/32/144, Annex I, *reprinted in Basic Documents*: 230-46.

28. The Treaty for the Prohibition of Nuclear Weapons in Latin America, *reprinted in Basic Documents*: 197-203.

29. The Treaty on the Non-Proliferation of Nuclear Weapons, *reprinted in Basic Documents*: 204-06.

30. For details, see, *e.g.*, John R. Redick, "Latin America: Reducing the Threat of Nuclear Proliferation," *Wisconsin International Law Journal* 5 (1987): 79.

31. For details, see, *e.g.*, Note, "Strengthening the Nuclear Non-Proliferation Regime: A Draft Convention," *Columbia Journal of Transnational Law* 26 (1987): 167.

32. The Commission is headquartered at 325 Ninth Street, San Francisco, CA 94103. Members of the Commission, in addition to the author, are Anders Boserup, William Epstein, Willis Harman, Donald Keys, Ward Morehouse, James O.C. Jonah, Betty Reardon, Indar Jit Rikhye, and Louis B. Sohn.

33. The Committee is headquartered at 225 Lafayette Street, Suite 207, New York, NY 10012.

34. The Institute is headquartered at 777 United Nations Plaza, New York, NY 10017.

35. *See* Robert C. Johansen, "Toward an Alternative Security System," in Burns H. Weston (ed.), *Toward Nuclear Disarmament and Global Security: A Search for Alternatives* (Boulder, CO: Westview, 1984): 569, 582-92.

36. *Ibid.*, pp. 584-85 (Table 1). *See also* the chapter by Bruce Russett in this volume — "Politics and Alternative Security: Toward a More Democratic and Therefore More Peaceful World" — wherein the author convincingly argues that greater political and economic democracy at the national level ("depolarization" in Johansen's terms) makes for more durable peace on the international plane.

37. *See* writings cited in note 5.

38. Many of the ideas enumerated here are derived from Richard A. Falk, "Toward a Legal Regime for Nuclear Weapons," *McGill Law Journal* 28 (1963): 519, 537-38.

39. The Trident II (D-5) missile is, concededly, a borderline case and also is not as destabilizing as the other missiles mentioned. Being by and large invulnerable to attack, it does not suffer the same time pressure to "use it or lose it" as do the other missiles.

40. *See* note 11.

41. *See, e.g.,* the Draft Treaty Banning Nuclear Weapons Tests in All Environments, *Department of State Bulletin* 47 (1962): 411.

42. *See* notes 30-31.

43. Treaty Between the United States of America and the Union of Soviet Socialist Republics on the Limitation of Anti-Ballistic Missile Systems, *reprinted in Basic Documents*: 213-15.

44. For pertinent discussion of these and related matters, see Francis A. Boyle, *Defending Civil Resistance Under International Law* (Dobbs Ferry, NY: Transnational Publishers, 1987).

45. *See* note 3.

46. For pertinent discussion, see Weston, "Nuclear Weapons Versus International Law."

47. Protocol for the Prohibition of the Use in War of Asphyxiating, Poisonous or Other Gases, and of Bacteriological Methods of Warfare, *reprinted in Basic Documents*: 136.

48. Convention on the Prohibition of the Development, Production and Stockpiling of Bacteriological (Biological) and Toxin Weapons and on Their Destruction, *reprinted in Basic Documents*: 211-12.

49. For related discussion, see this chapter: A NATO-Warsaw Pact Nonaggression Regime. *See also* the chapter by Thomas F. Lynch, III, in this volume.

50. *See, e.g.,* Ruth Leger Sivard, *World Military and Social Expenditures 1987-88*, at 10-13 (Washington, DC: World Priorities, Twelfth Edition, 1987).

51. For details, see Michael Klare, *American Arms Supermarket* (Austin, TX: University of Texas Press, 1984).

52. George F. Kennan, *The Nuclear Delusion* (New York: Pantheon Books, 1982): xxviii.

53. Arguably such a non-aggression regime would have to be accompanied by substantial mutual disarmament between the military superpowers.

54. Interest recently has been reawakened in the 1967 "Rapacki Plan" (named after then Polish foreign minister Adam Rapacki) pursuant to which the central front of Europe would have been demilitarized by removing all nuclear weapons from Czechoslovakia, the two Germanys, and Poland as well as most conventional forces from a 300-mile corridor along the German border. In May 1987, Poland introduced a new plan calling for the "gradual and mutually agreed withdrawal" of both nuclear and conventional weapons from Czechoslovakia, East Germany, Hungary, and Poland within the Warsaw Pact, and from Belgium, Denmark, Luxembourg, the Netherlands, and West Germany within NATO. *See, e.g.,* Michael T. Kaufman, "Polish Chief Offers Plan for Arms Disengagement," *N.Y. Times*, May 9, 1987: 3.

55. Proposals for non-offensive or non-provocative defense are not new, least of all in Europe where debate on the subject has been going on for nearly a decade and a half. The basic principle is simple: to prevent attack by a potential adversary the potential adversary must be made to understand (1) that an act of aggression will incur severe retaliation and (2) that there is nothing to fear from the potential target State in the absence of an act of aggression. *See, e.g.,* Dietrich Fischer, *Preventing War in the Nuclear Age* (Totowa, NJ: Rowman and Allanheld, 1989); Dietrich Fischer, Wilhelm Nolte,

and Jan Oberg, *Winning Peace* (New York: Crane, Russak, 1989). *See also* Hollins, Powers, and Sommer, *The Conquest of War:* 78-88; Dietrich Fischer, "Invulnerability Without Threat: The Swiss Concept of General Defense," in Burns H. Weston (ed.), *Toward Nuclear Disarmament and Global Security: A Search for Alternatives* (Boulder, CO: Westview, 1984).

56. For wide-ranging discussion of this and related human rights concerns, see Richard P. Claude and Burns H. Weston (eds.), *Human Rights in the World Community: Issues and Action* (Philadelphia: University of Pennsylvania Press, 1989).

57. There is a growing body of literature detailing the various forms that a non-intervention regime might take, globally and regionally. *See, e.g.,* Randall Forsberg, "The Case for a Third World Nonintervention Regime," *Alternative Defense Working Paper No. 6* (Brookline, MA: Institute for Defense and Disarmament Studies, December 1987). *See also* Robert O. Keohane, *After Hegemony: Cooperation and Discord in the World Political Economy* (Princeton, NJ: Princeton University Press, 1984); Stephen D. Krasner (ed.), *International Regimes* (Ithaca, NY: Cornell University Press, 1983).

58. Some of the ideas enumerated here are derived from the *Report of the Independent Commission on Disarmament and Security Issues: Common Security—A Programme for Disarmament*, also known as the "Palme Commission Report," U.N. Doc. A/CN.10/38 (1983).

59. The ideas enumerated here are derived in part from Louis B. Sohn, *Peaceful Settlement of Disputes and International Security*, a preliminary draft of an unpublished manuscript submitted to the Independent Commission on World Security Alternatives (*see* note 32).

60. *See* Richard A. Falk, *Reviving the World Court* (Charlottesville, VA: University Press of Virginia, 1986). *See also* Thomas M. Franck, *Judging the World Court* (New York: Priority Press, Twentieth Century Fund, 1986).

61. *See* this chapter: Institutional Policy Options.

62. *See* page 83.

63. For the first two proposals enumerated here, I am indebted in part to Daniel Arbess and William Epstein, "Disarmament Role for the United Nations?" *Bulletin of the Atomic Scientists* 41 (May 1985): 26, 28.

64. Such an agency—an international satellite monitoring agency (ISMA) —was proposed by a special United Nations commission in 1982 to monitor arms control agreements and perform related other functions. Although the plan was ultimately blocked by the superpowers, interest in variations on it have grown ever since.

65. *See* note 11.

66. *See* this chapter: Improvement of Opportunities and Capabilities for Legal Challenges to Coercive Foreign Policies.

67. This recommendation was stimulated by Rajni Kothari *et al.*, *Towards a Liberating Peace* (New York: New Horizons Press, 1988): 47-48, a product of the United Nations University's Programme on Peace and Global Transformation.

68. For further discussion along these lines, see the chapter by Lloyd J. Dumas in this volume and Lloyd J. Dumas, *The Political Economy of Arms Reduction: Reversing Economic Decay* (AAAS Selected Symposium No. 80, 1982). *See also* Howard S. Brembeck, *The Civilized Defense Plan: Security of Nations Through the Power of Trade* (Fairfax, VA: Hero Books, 1989).

69. *See, e.g.,* Robert C. Johansen and Saul H. Mendlovitz, "The Role of Law in the Establishment of a New International Order: A Proposal for a Transnational Police Force," *Alternatives: A Journal of World Policy* 6 (1980): 307.

70. *See, e.g.,* the Draft Statute for an International Criminal Court prepared under the auspices of the U.N. Commission on International Criminal Jurisdiction in 1953, 9 U.N. GAOR, Supp. (No. 12) Annex p. 23, U.N. Doc. A/2645 (1954). *See also* Julius Stone, *An International Criminal Court* (World Peace Through Law Conference, Geneva, 1971); Fannie L. Klein and Daniel Wilkes, "United Nations Draft Statute for an International Criminal Court: An American Evaluation," in Gerhard O. W. Mueller and Edward M. Wise (eds.), *International Criminal Law* (South Hackensack, NJ: F. B. Rothman, 1965): 526.

71. *See,* for example, the proposals recommended in this chapter: Improvement of UN Peacekeeping Opportunities and Capabilities. *See also* Inis L. Claude, *Swords Into Plowshares* (New York: Random House, Fourth Edition, 1971): chs. 4 (*Collective Security: An Alternative to Balance of Power?*) and 5 (*A Critique of Collective Security*); Chadwick Alger, "Reconstructing Human Polities: Collective Security in the Nuclear Age," in Burns H. Weston (ed.), *Toward Nuclear Disarmament and Global Security: Search for Alternatives* (Boulder, CO: Westview, 1984): 666.

72. *See, e.g.,* Silviu Brucan, "The Establishment of a World Authority: Working Hypotheses," in Burns H. Weston (ed.), *Toward Nuclear Disarmament and Global Security: A Search for Alternatives* (Boulder. CO: Westview, 1984): 615. *See also* Richard A. Falk, *A Study of Future Worlds* (New York: The Free Press, 1975).

73. For pertinent discussion, see, *e.g.,* Patricia M. Mische, "Ecological Security and the Need to Reconceptualize Sovereignty," *Alternatives* 14, 4 (October 1989): 389. *See also* Jessica Tuchman Matthews, "Redefining Security," *Foreign Affairs* 68 (1989): 162.

74. That environmental decline can lead to conflict and potentially violence albeit not as self-evidently as social injustice and economic malaise because it does so more indirectly, is noted in Matthews, "Redefining Security": 166, "Environmental decline occasionally leads directly to conflict, especially when scarce water resources must be shared. Generally, however, its impact on nations' security is felt in the downward pull on economic performance and, therefore, on political stability. The underlying cause of turmoil is often ignored; instead governments address the poverty and instability that are its results."

75. For insightful confirmation of this proposition, see Kothari, *et al., Towards a Liberating Peace: passim.*

76. For pertinent discussion, see Burns H. Weston, Richard A. Falk, and Anthony D'Amato (eds.), *International Law and World Order: A Problem-Oriented Coursebook* (St. Paul, MN: West Publishing, Second Edition, 1990): Part III.

77. Pierre Teilhard de Chardin, *The Future of Man* (New York: Harper and Row, First American Edition, Norman Denny translation, 1964): 115.

78. Loren Eiseley, *The Immense Journey* (New York: Vintage Books, 1959): 140.

4

Politics and Alternative Security: Toward a More Democratic, Therefore More Peaceful, World

Bruce Russett

A government that will break faith with its own people cannot be trusted to keep faith with foreign powers.
— Ronald W. Reagan, 1986

In this Moscow spring, this May 1988, we may be allowed to hope that freedom . . . will blossom forth at last in the rich soil of your people and culture. We may be allowed to hope that the marvelous sound of new openness will keep rising through, ringing through, leading to a new world of reconciliation, friendship, and peace.
— Ronald W. Reagan, 1988

We will do the worst thing to you — we will deprive you of your enemy.
— A. Georgi Arbatov, 1987

Two apparent facts about contemporary international patterns of war and peace stare us in the face. The first is that some States expect, prepare for, and fight wars against other States. The second is that some States do *not* expect, prepare for, or fight wars *at least against each other.* The first is obvious to everyone. The second is widely ignored, yet it is now true on an historically unprecedented scale, encompassing wide areas of the earth. In a real if still partial sense, peace is already among us. We need only recognize it, and try to learn from it.

An understanding of why some States do not engage in hostility may lead us to an attainable basis for an alternative system of security, one that does not depend on acceptance of a world State to enforce peace or on a particular configuration of strategy and weaponry to provide a peace of sorts through some form of stable deterrence. Accordingly, in this chapter I will ignore many of the commonly known (and partly useful) institutional and military proposals for alternative security, and instead explore the causes, limitations, and implications of this political anomaly of limited peace already among us.

In so doing, I cannot but acknowledge my own historical grounding, including that of a privileged member of society in a powerful capitalist country, governed by democratic procedures as understood in the western liberal tradition. That context must inevitably color both my discourse on the observation of "facts" and my normative evaluation of their implications. Persons situated in other contexts may well dispute either or both. My perspective on both scientific statements and human rights is one of moderate historicism; that whereas all are in some sense conventions, they can be substantially grounded across ages and cultures.[1] I hope here to demonstrate the reasons, embedded in contemporary social practice, for an emerging consensus in interpreting some highly important experiences, and to contribute to the development of a disciplined, if necessarily more tenuous, agreement on their evaluation.

PEACE AMONG DEMOCRACIES

I am referring to the peace among the industrialized and democratically governed States, primarily in the northern hemisphere. These States – members of the Organization for Economic Cooperation and Development (OECD: Western Europe, North America, Japan, Australia, and New Zealand), plus a few scattered less industrialized democratic States – constitute a vast zone of peace, with more than three quarters of a billion people. Not only has there been no war among them for almost 45 years, as shown in Table 4.1, there has been little expectation of, or preparation for, war among them either. By war I mean large-scale

TABLE 4.1
Distribution of International Wars, 1945-88

	Fought in		
Fought by	OECD countries	Communist countries	Less developed countries
OECD countries	0	1	7
Communist countries	0	3	3
Less developed countries	0	1	19

Source: Melvin Small and J. David Singer, *Resort to Arms: International and Civil Wars* (Beverly Hills, CA: Sage, 1982), updated to 1988. Includes all interstate and colonial wars (not civil wars) with over 1,000 battle deaths.

organized international violence with at least 1,000 battle deaths (by a conventional social science definition). In fact, even much smaller-scale violence between these countries has been virtually absent. The nearest exception is Greece and Turkey, especially with their brief and limited violent clashes over Cyprus. They are, however, among the poorest countries of this group and only sporadically democratic.

In the years before 1945, many of them fought often and bitterly, but always when at least one of the States in any warring pair was ruled by an authoritarian or totalitarian regime. Despite that past, war among them is now virtually unthinkable. What seemingly had been the most permanent enmities—for instance, between France and Germany—have for the past two or three decades appeared well buried. Individual citizens may not love each other across national boundaries, but neither do they expect the other's State to attack nor do they wish to mount an attack on it. Expectations of peace are thus equally important; these peoples make few preparations for violence between them; peace for them means more than just the prevention of war through threat and deterrence. This condition has been characterized as a "security community" or as "stable peace."[2] By the standards of world history this is an extraordinary achievement.

It is not easy to explain just why this peace has occurred. Partly it is due to the network of *international law and institutions* deliberately put into place to make a repetition of the previous world wars both unthinkable and impossible. But that network is strongest in Western Europe, often excluding the countries in North America and the Far East; even in the strongest instance the institutions typically lack full powers to police and coerce would-be breakers of the peace — and, as we shall see, even powerful institutions alone cannot guarantee peace if the underlying preconditions of peace are lacking.

In part it is due to favorable economic conditions associated with advanced capitalism. Fairly steady economic growth, a high absolute level of prosperity, relative equality of incomes within and across the industrial States, and a dense network of trade and investment across national borders all make resort to violence dubious on cost-benefit grounds; a potential aggressor who already is wealthy risks much by resorting to violence for only moderate gain.[3] But the condition of peace among these rich States has not been endangered by periods of postwar recession and stagnation, and in other parts of the world, especially Latin America, there are democratic States that are not wealthy but still at peace with one another.

Partly too, peace is the result of a perceived external threat faced by the industrialized democracies. They maintain peace among themselves in order not to invite intervention by the communist powers. Where peace among them is threatened, it may be enforced by the dominant hegemonic power of the United States. But the external threat also has waxed and waned without affecting the peace among these States. Indeed, their peace became even more stable during the 1970s and 1980s, the very time when the Cold War abated and Europeans, especially, ceased to have much fear of Soviet attack. All these explanations, therefore, are at best only partial ones, and we are driven back to observing that the period of peace among the highly industrialized States essentially coincides with the period when they all have been under democratic rule.[4]

Conceptually and empirically, the competing explanations overlap somewhat and reinforce one another, especially for the post-World War II era. International law has served to legitimate widely many of the domestic legal principles of human rights associated with liberal democracy; all the capitalist industrial States have been, since World War II, democratic (though not all democratic States are economically advanced); most of them have also been part of the US hegemonic alliance system (which has included also non-democratic and economically less developed countries). Although this overlap prevents a definitive test, all the alternative hypotheses find their predictions falsified by at least one warring pair: *i.e.*, the British-Argentine war in 1982, between two capitalist (Argentina only moderately advanced) States allied with the United States. World Wars I and II, of course, included many industrial capitalist countries as warring pairs. Analysts as different as Joseph Schumpeter and Karl Kautsky predicted peace among advanced capitalist States; Lenin did not. Nor is it simply part of a general statement that politically or culturally similar countries do not fight each other.[5] An empirical correlation between cultural similarity and relative absence of war exists, but it is a weak one. There are several examples of wars or threats of war within Eastern Europe and Latin America in recent decades; by contrast, a reduction in regional enmities is associated with parallel democratization (*e.g.*, Argentina and Brazil).

Another reason to doubt explanations relying chiefly on international institutions, economic conditions, or external threat is that the experience of peace among democratic countries goes back (among fewer countries, to be sure) at least to the end of the Napoleonic wars in 1815. Previous records are less precise, but also less relevant, since democracy as we know it in this era was at best a rarity before then. In ancient Greece, Athens and Sparta typically allied with democracies and oligarchies respectively. They often intervened to change the domestic constitution of allies to their preferred mode; similarly, a change in domestic constitution among the smaller city-states often produced switches in alliances. Athens did, however, occasionally attack democratic cities, as in Sicily.[6]

With only the most marginal exceptions, democratic States have not fought each other in the modern era. This is perhaps the strongest non-trivial or non-tautological statement that can be made about international relations. The nearest exception is Lebanon's peripheral involvement in Israel's "War of Independence" in 1948. Israel had not yet held an election, so Melvin Small and J. David Singer did not count it as a democracy.[7] Other exceptions are truly marginal: in 1849 between two States both briefly democratic (France and the Papal States), and Finland against the western allies in World War II (nominal only because Finland's real quarrel was with the USSR). In the War of 1812 with the United States, Britain's franchise was sharply restricted, as was the Boer Republic's in its attempt to preserve its independence against Britain in 1898.

By democratic State I mean the conditions of public contestation and participation, essentially as identified by Robert Dahl, with a voting franchise for a substantial fraction of male citizens (in the Nineteenth and early Twentieth centuries and wider thereafter), contested elections, and an executive either popularly elected or responsible to an elected legislature.[8] While scholars who have found this pattern differ slightly in their definitions, agreement on the condition of non-war among democracies ("liberal," "libertarian," or "polyarchic" States) is now overwhelming.[9] This simple fact cries out for explanation: what is it about democratic government that inhibits people from fighting one another?

In exploring that question we should be clear about what is not implied. The condition of peace *between* democratic States does *not* mean that democratic States are *ipso facto* peaceful with *all* countries. They are not. In their relations with non-democratic States—whether great powers, weak States, or non-western peoples essentially outside the State system and hence "available" as targets for imperial expansion—they have often fought, more or less as frequently as non-democratic States have fought or prepared to fight (*see* this chapter: Relations with Non-Democratic States). This distinction has been widely noted, and though there is some dissent the scholarly consensus is nearly complete.[10]

INTERNAL PEACE AND INTERNATIONAL PEACE

There are powerful norms against the use of lethal force both within democratic States and between them. Within them is of course the basic norm of liberal democratic theory—that disputes can be resolved without force through democratic political processes, which in some balance are to ensure both majority rule and minority rights. A norm of equality operates both as voting equality and as certain egalitarian rights to human dignity. Democratic government rests on the consent of the governed, but justice demands that consent not be abused. Resort to organized lethal violence, or the threat of it, is considered illegitimate and unnecessary to secure one's legitimate rights. Dissent within broad limits by a loyal opposition is expected and even needed for enlightened policy-making, and the opposition's basic loyalty to the system is to be assumed in the absence of evidence to the contrary.

All participants in the political process are expected to share these norms. In practice, the norms do sometimes break down, but the normative restraints on violent behavior—by State and citizens—are fully as important as the State's monopoly on the legitimate use of force in keeping incidents of the organized use of force rare. In fact, the norms are probably more important than is any particular institutional characteristic (two party/multi-party, republican/parliamentary) or formal constitutional provision.   Institutions may precede the development of norms.   If they do, the basis for restraint is likely to be less secure.

Democracy did not suddenly emerge full-blown in the West; nor did it emerge by any linear progression. Only over time did it come to mean the extension of a universal voting franchise, formal protection for the rights of ethnic, racial, and religious minorities, and the rights of groups to organize for economic and social action. The rights to organize came to imply the right to carry on conflict— non-violently—as by strikes, under the principle that both sides in the conflict had to recognize the right of the other to struggle, so long as that struggle was constrained by law, mutual self-interest, and mutual respect. The implicit or explicit contract in the extension of such rights was that the beneficiaries of those rights in turn would extend them to their adversaries.

To observe this is not to accept democratic theory uncritically, nor to deny that it is part of a belief structure that, in Antonio Gramsci's view of cultural hegemony, may serve to legitimate dominant class interests and provide subordinate classes with a spurious sense of their own political efficacy.[11] As such, it may exaggerate belief in the reasonableness both of one's own State's demands in international politics and those of other democratic States. But it is precisely beliefs and perceptions that are primarily at issue here; insofar as the other State's demands are considered *ipso facto* reasonable according to a view of one's own system that extends to theirs, popular sentiment for war or resistance to compromise is undermined.

Politics within a democracy is seen as largely a non-zero sum enterprise: by cooperating, all can gain something even if all do not gain equally, and the winners today are restrained from crushing the losers; indeed, the winners may, with shifting coalitions, wish tomorrow to ally with today's losers. If the conflicts degenerate to physical violence, either by those in control of the State or by insurgents, all can lose. In international politics—the anarchy of a self-help system with no superordinate governing authority—these norms are not the same. "Realists" remind us of the powerful norms of legitimate self-defense and the acceptability of military deterrence, norms much more extensive internationally than within democratic States. Politics between nations takes on a more zerosum hue. True, we know we all can lose in nuclear war or in a collapse of international commerce, but we worry much more about comparative gains and losses. The essence of realist politics is that even when two States both become more wealthy, if one gains much more wealth than the other it also gains more power potential to coerce the other; thus the one that is lagging economically only in relative terms may be an absolute loser in the power contest.

The principles of anarchy and self-help in a zero-sum world are seen most acutely in "structural realist" theories of international relations. Specifically, a bipolar system of two great States or alliances, each much more powerful than any others in the international system, is seen as inherently antagonistic. The nature of the great powers' internal systems of government is irrelevant; whatever they may work out with or impose on some of their smaller allies, their overall behavior with other great powers is basically determined by the structure of the international system and their position in that structure. Athens and Sparta, the United States and the Soviet Union, all are doomed to compete and to resist any substantial accretion to the other's power. To fail to compete is to risk the death of sovereignty, or death itself. Through prudence and self-interest they may avoid a war that might destroy or cripple both of them (the metaphor of two scorpions in a bottle); but the threat of war is never absent and never can be absent. Peace, such as it is, can come only from deterrence, eternal viligance, and probably violent competition between their proxies elsewhere in the world. By this structural realist understanding, the kind of stable peace that exists between the democratic countries never can exist on a global scale.[12]

Efforts to establish norms against the use of lethal violence internationally have been effective only to a limited degree. The Kellogg-Briand Pact of 1928 to outlaw war was a failure from the outset, as have been efforts to outlaw "aggressive" war. Despite its expression of norms and some procedures for the pacific settlement of disputes, the UN Charter fully acknowledges "the inherent right of individual or collective self-defense if an armed attack occurs" (Article 51). It could hardly do otherwise in the absence of superordinate authority. The norm of national self-defense — including collective self-defense on behalf of allies, and defense of broadly conceived "vital" interests even when national survival is not at stake — remains fully legitimate to all but tiny pacifist minorities. Although there is some cross-cultural variation in the readiness of different peoples to use lethal force in different modes of self-defense, these differences are not strongly linked to form of government. Citizens of small democracies who perceive themselves as beleaguered (*e.g.*, Israel) or citizens of large powerful democracies with imperial histories or a sense of global responsibility for the welfare of others (*e.g.*, Britain, the United States) are apt to interpret national or collective interest quite broadly. Especially across international cultural barriers, perversions of the right of self-defense come easily.

Yet democratic peoples exercise that right within a sense that somehow they and other peoples *ought* to be able to satisfy common interests and work out compromise solutions to their problems, without recourse to violence or threat of it. After all, that is the norm for behavior to which they aspire within democratic systems. Because other people living in democratic States are presumed to share those norms of live and let live, they can be presumed to share their moderate behavior in international affairs as well. That is, they can be respected as self-governing peoples and expected to offer the same respect to other democratic countries in turn.[13] The habits and predispositions they show in their behavior in internal politics can be presumed to apply when they deal with like-minded outsiders. If one claims the principle of self-determination for oneself, normatively one must accord it to others perceived as self-governing. Norms do matter. Within a transnational democratic culture, as within a democratic nation, others are seen as possessing rights and exercising those rights in a spirit of enlightened self-interest. Acknowledgment of those rights both prevents us from wishing to dominate them and allows us to mitigate our fear that they will try to dominate us.

Realism has no explanation for the fact that certain kinds of States— namely, democratic ones—do not fight or prepare to fight one another. One must look instead to the liberal idealist vision of Immanuel Kant's *Perpetual Peace*, embodied also in Woodrow Wilson's vision of a peaceful world of democratic States. This same vision inspired US determination to root out fascism and establish the basis for democratic governments in West Germany and Japan after World War II (and partly also explains, and was used to justify, interventions in Vietnam, Grenada, Nicaragua, and so on).

Democratic States, with their wide variety of active interest groups in shifting coalition patterns, also present the opportunity for transnational coalition formation in alliance with groups in other democracies. This may seem a form of "meddling"; it also provides another channel for international conflict resolution. International anarchy is not supplanted by institutions of common government, but conflicts of interest within the anarchy can be moderated fairly peacefully on the principle of self-determination within an international society.

How much importance should we attribute to perceptions among the public in general, and how much to those of the elites including, in particular, the leaders of the State? Decisions for war, and indeed most major decisions in national security matters, are taken by the leaders and debated largely among the elites. They have substantial ability to lead and mold mass opinion. The public can be expected to support most foreign policy acts, including uses of military force, at least in the short run – a phenomenon widely understood as the rally-round-the-flag effect. Elite views of the world are typically more differentiated than are those of the general public; the elites have more information and distinguish more sharply in their likes and dislikes. In periods of relatively favorable attitudes toward the Soviet Union, Americans with higher income and social status are most likely to hold those favorable attitudes.[14] Elite readiness to view other peoples as self-governing should matter most.

Nevertheless, the elites in a democracy know that the expenditure of blood and treasure in any extended or costly international conflict can be sustained only with the support of the general public. Whereas there may be leads and lags either way, long-term serious differences between public opinion and official foreign policy are rare.[15] Hence the elites will be somewhat constrained by popular views of the reasonableness of engaging in violent conflict with a particular foreign country.

In some ways the principle of self-determination may actually work better in the absence of a common government. If there were a set of central institutions for common government, different groups and peoples would by necessity compete to control it, with the risk that control (majority rule) would be abused at the expense of minority rights. A common government would have the legal right and powers to tax, to coerce, and to re-allocate wealth and benefits. For the institutions to work peaceably, the norms must be strong and widely shared. In the absence of broad agreement on politics and culture, it is best that the institutions, and the possibility for their abuse, also be absent. Hence, some peoples can live with each other peaceably under separate governments but not under a common one. (Contrast, for example, relations between Protestant Britain and Catholic Ireland with those between Protestants and Catholics within Northern Ireland.) The formal institutions of democratic government might be in place under a common State, but the degree of sharing in the norms of self-restraint, and confidence that others share those norms, would be inadequate to ensure peace. Hence comes also the common fear of a world leviathan containing very diverse peoples, *even* under some form of direct election and representation. The norms might well be insufficient to restrain action, especially given the extreme economic inequalities of the contemporary global community.

RELATIONS WITH NON-DEMOCRATIC STATES

When we look within the construct of democratic ideology, it is apparent that the restraints on behavior that operate between separately governed democratic peoples do not apply to their relations with non-democratic States. If other self-governing (democratic) peoples can be presumed to be worthy of being treated in a spirit of compromise and, in turn, to be acting in that spirit, the same presumption does not apply to authoritarian States. According to democratic norms, authoritarian States do not rest on the proper consent of the governed, and thus they cannot properly represent the will of their peoples (if they did, they would not need to rule through undemocratic, authoritarian institutions). Rulers who control their own people by such means, who do not behave in a just way that respects their own people's right to self-determination, cannot be expected to behave better toward peoples outside their jurisdictions. "Because nonliberal governments are in a state of aggression with their own people, their foreign relations become for liberal governments deeply suspect. In short, fellow liberals benefit from a presumption of amity; nonliberals suffer from a presumption of enmity."[16] Authoritarian governments are expected to aggress against others if given the power and opportunity. By this reasoning, democracies must be eternally vigilant against them and may even sometimes feel the need to engage in preemptive or preventive (defensively motivated) war against them.

Whereas wars against other democratic States are neither expected nor really legitimate, wars against authoritarian States may often be expected and "legitimated" by the principles above. Thus, an international system composed of both democratic and authoritarian States will include both zones of peace (actual and expected, among the democracies) and zones of war or, at best, deterrence between democratic States and authoritarian ones and, of course, between authoritarian States. A condition of non-war may exist between States even if one of them is not a democracy, but only based on the power of one or both States to deter the other from the use of lethal force: the one-sided deterrence of dominance or mutual deterrence between those more or less equally powerful. If the democratic State is strong, its forbearance may permit war to be avoided.

Of course, democracies have not fought wars only out of motivations of self-defense, however broadly one may define self-defense to include "extended deterrence" for the defense of allies and other interests or to include anticipation of others' aggression. Many of them have also fought imperialist wars to acquire or hold colonies or, since World War II, to retain control of States formally independent but within their spheres of influence. In these cases they have fought against people who on one ground or another could be identified as not self-governing.[17]

In the days of Nineteenth Century colonial expansion, the colonized peoples were in most instances outside the European State system. They were in most instances not people with white skins. And they were in virtually every instance people whose institutions of government did not conform to the Western democratic institutional forms of their democratic colonizers. Europeans' ethnocentric views of those peoples carried the *assumption* that they did not have institutions of self-government, that their governments or tribal leaders were assumed not to be just or consensual. They were not merely available as candidates for imperial aggrandizement, they could also be considered candidates for betterment and even "liberation"—the white man's burden, or *mission civilatrice*. Post-Darwinian ideology even regarded them as at a lower stage of physical evolution and intellectual capacity than whites (and especially white males).[18] They could be brought the benefits not only of modern material civilization but of Western principles of self-government and, after proper tutelage, of the institutions of self-government. If they did not have such institutions already, then by definition they were already being exploited and repressed. Their governments or tribal leaders could not, in this ethnocentric view, be presumed to be just or consensual, and thus one need have few compunctions about conquering them. They were legitimate candidates for "liberal" imperialism.

Later, when western forms of self-government did begin to take root at least on a local basis in the colonies, the extremes of pseudo-Darwinism racism lost their legitimacy. As these things happened, the colonial powers' legitimacy in controlling those peoples were eroded. Indeed, indigenous leaders vigorously turned back onto their colonial rulers their very own principles (*e.g.*, independence leaders in the Philippines; or Gandhi and the Congress Party in India, who were especially effective normatively against a British Labour government deeply committed to providing equality at home). Decolonization came not only because the colonial governments had lost the power to retain their colonies but also because in many cases they lost confidence in their normative right to rule. The evolution of the colonies themselves – and of the understandings about colonial peoples that were held in the imperial states – eroded the legitimacy of the colonial rulers in their own eyes. The imperial peoples' liberal principles were turned back on them. In a further round, those principles are now being turned against Third World authoritarian rulers.[19]

Another important qualification must take account of a weakness of all peoples, that of scapegoating. It is an old trick to blame outsiders – either socially marginal groups within a country, or external adversaries – when things go wrong. It is often popular to attribute troubles to foreigners or to try to turn people's frustrations against external enemies, whether or not those outside can plausibly be blamed for the troubles. This kind of behavior has long been attributed to dictatorships, with examples including Nazi persecution of Germany's Jews and, more recently, the Argentine junta's decision, in 1982, to choose a time of economic stagnation and political unrest to stoke popular nationalism and seize the Falklands/Malvinas islands. Ironically, however, certain forms of scapegoating behavior may be equally prevalent in democracies. British Prime Minister Margaret Thatcher was also unpopular with her electorate and used the Argentine attack as a very effective rallying point toward a sweeping election victory a year later.[20]

Leaders of the United States, for example, have been more likely to use military force or to pursue confrontational policies against foreign adversaries when the US economy has been in stagnation or recession. This pattern emerges both in short-term analyses of events since World War II and in long-term examinations of the past century. Furthermore, confrontational policies have been somewhat more common just before elections, and uses of force or threats to use force most commonly engaged in, over the century, in years when elections and recent economic adversity were both present.[21]

The president or prime minister of a democracy faces enormous demands and expectations for the provision of prosperity, security, and all the other elements of the good life. Expression of these demands is encouraged by the personalization of politics (the decline of parties as institutions mediating between leaders and people) and modern means of telecommunication, which bring both problems (recession, foreign conflicts) and the face and voice of the leader directly into everyone's living room. The demands far exceed the capabilities of any political leader under most circumstances; hence the leader is often reduced to trying to give the appearance of solving them, through rhetoric or symbolic measures with at best short-term effects or no real effect at all (consider, for example, iterations of the "war on drugs"). "The desperate search [of the President on behalf of his public] is no longer for the good life but for the most effective presentation of appearances. This is a pathology because it escalates the rhetoric at home, ratcheting expectations upward notch by notch, and fuels adventurism a-broad."[22] Talking tough to the Soviet Union or using military force against a weaker adversary (*e.g.*, Grenada or Libya) can be a tempting way to produce at least a temporary boost in the leader's popularity through what has become widely known as the rally-round-the-flag effect.[23]

Democratic governments may therefore often play on, encourage, and pander to popular nationalism; they may do so even more than authoritarian governments just because democratic leaders are often less secure in their offices. They know they must face another election within a very few years and meanwhile must project an aura of electoral popularity as an instrument to persuade reluctant legislators to support their programs. They do not necessarily intend war. Talking or acting tough typically is popular because it is seen as preventing war by deterrence. War itself usually is not popular, especially to the degree that it becomes protracted or expensive in wealth or blood.[24] But insofar as there can be found weaker States who can be bullied or attacked in relative safety, governed by rulers who are plausibly not democratic and therefore unrepresentative of their peoples' "real" wishes, the incentive to belligerent behavior toward other states is substantial even in democracies. It may be strengthened by the very virtue of democracy in giving to the mass public a degree of control over its fate: mass opinion is typically less informed and in some real sense more ethnocentric and less cosmopolitan, than elite opinion. That ethnocentricism may be magnified when confronted by conflicts with other peoples who are not governed "like us."

A SHIFT TOWARD DEMOCRACY

The end of World War II brought in its wake the demise of colonial empires; it brought a degree of self-determination to the formally colonized peoples. Unfortunately, that self-determination was often highly restricted, limited in part by ties of economic and military neo-colonialism. Self-determination also was often limited to the elites of the new States, as the governments installed were frequently authoritarian and repressive—anything but democratic. Yet there has been, since the mid-1970s, some evolution toward greater democracy in large parts of what is called the Third World. In 1973 only two Spanish- or Portuguese-speaking States in South America were governed by democratic regimes (Colombia and Venezuela); now only two are ruled by military dictatorships (Paraguay and, in transition, Chile). Democracy remains fragile and imperfect in many of them,[25] but the relative shift away from authoritarian rule is palpable.

This shift shows up statistically on a worldwide basis. A long-term observer of political rights and civil liberties, Raymond Gastil, has carried out, over this period, a project of rating countries according to their degree of "political freedom."[26] His rating is not meant to reflect a broad definition of human rights that includes, for example, the so-called second generation economic rights to employment or the satisfaction of basic physical needs. Rather, it addresses first generation rights: electoral practices, the accountability of the executive and legislature, judicial procedures, and freedom of expression and association—in short, dimensions of the traditional political definition of democracy.[27] For some of his purposes, Gastil uses two scales of seven points each; for others he collapses these complex judgments into three categories of States: free, partly free, and not free. The distribution of States in these three categories has varied over time as shown in Table 4.2.

By this evaluation, there has been a substantial decline in the number of strongly authoritarian (not free) States and a similarly substantial increase in the number of partly free States, especially if 1973 is used as the base year.[28] These trends can be seen in many parts of the world: the demise of dictatorships in Greece, Portugal, and Spain in Europe; Gastil's recent characterization of Hungary, Poland, and even the Soviet Union with scores in the "partly free" range; improvements he noted in China (as of 1988, and still overall "not free"); and shifts in several large, important countries elsewhere,

TABLE 4.2
World Percentage Distribution of States by Degree
of Political Freedom, 1973, 1976, and 1988

	1973	1976	1988
Free	32	30	35
Partly free	24	32	32
Not free	44	39	33
Number of States	165	165	165

Source: Raymond Gastil, *Freedom in the World: Political Rights and Civil Liberties, 1988-89* (New York: Freedom House, 1989).

to wit, Argentina, Brazil, India (since 1976 and its "emergency rule"), Pakistan, Philippines, South Korea, Thailand, and Uganda.

The degree of democratization should not be exaggerated, and the most substantial increase is only in the "partly free" category. The first shifts—to popular access to alternative sources of information and relative freedom of expression—are the easiest and most costly for governments to suppress. Pressure to cross further thresholds of democratization—the development of alternative political organizations, and of free, fair elections—represent greater threats to governmental power; they may seem to follow inexorably and yet at the same time meet with stiffer resistance by the State.[29] Whatever the ultimate outcome, recent developments may indicate more than just a cyclical alternation of democracy and dictatorship in an extension of the fluctuations of the 1930s and early 1970s. Rather, they may be an extension of a very long-term trend of global norms produced by the succession, since the seventeenth century, of global leaders with increasingly democratic internal political systems (the Dutch Republic, Britain, and the United States).[30] Some influences operating in the Philippines case of 1986 may suggest similar conditions that can reinforce movements toward democracy elsewhere:

1. the "demonstration" or "contagion" effect of the restoration of democracy in a number of States, especially in Latin America, sharing important cultural similarities with the Philippines;

2. belated but still effective US political intervention against continuation of the Marcos regime;
3. the role of national and international television – cleverly exploited by the revolutionaries who made seizure of the television station a prime objective – which brought the full glare of publicity onto any violent government suppression of the demonstrators (globally-observed bloodshed would have further undermined Marcos' already fading legitimacy; yet his failure to order violent suppression of the demonstrators permitted their success);
4. the role of international organizations in protecting, deliberately or otherwise, key centers of opposition to this regime, Cardinal Jaime L. Sin's position being especially critical;[31] and
5. the role of expatriates (especially Filipinos in Hawaii and the mainland United States) in providing experience and financial support to the opposition.

Obviously these influences do not apply equally to all cases, but one can observe them in lesser degrees in places as different as Poland and South Korea. All reflect instruments by which developing international norms about political rights can be made effective.

Arguably the shift to democracy is substantially a result of the manifest economic as well as political failures of dictatorships: authoritarian regimes instituted in the name of economic growth and national development that were in fact unable to deliver on their promises. Perhaps after a spurt of economic growth, they all too often brought stagnation, greater economic inequality, and a loss of true national autonomy, in addition to the suppression of political liberties. It is no wonder that they lost favor with their peoples. Democratic governments also may ultimately lose favor if they are unable to revive stagnant economies – an especially severe danger in Latin America, though many peoples have so far shown a good deal of tolerance for their governments' predicaments. Economic failures, in communist and capitalist States, might ultimately increase support for nationalist, fundamentalist, or fascist ideologies rather than for democracy. For now we can perhaps be permitted a degree of hopefulness and even an assertion that authoritarian rule is out of favor in the global culture, with effects in communist countries and in the Third World.

It can always be a cause for rejoicing when people gain more power over their own fate, with a widening and deepening of the institutions and practices of democratic government. Does this analysis imply something more, that if the shift toward democracy does continue, we would then move into an era of international peace? If all States were democratic, could we all live in perpetual peace? Does a solution lie in creating a world where all countries are governed by democratic practices? In principle, this would both rid the world of aggressive behavior of some kinds of autocratic regimes and deprive democratically governed peoples of a normatively legitimate target for jingoism.

A serious reservation, however, must concern interpretation of the word "creating." The argument here does *not* imply that the route to ultimate perpetual peace is through wars, or threats of war, to make other countries democratic. External threats all too commonly become means to reinforce, not relax, the repressive power of the State. Wars are corrupting to those who fight them, serving to legitimate violence within as well as between countries.[32] The self-righteous temptation to blame the adversary and dehumanize the enemy is too strong to give any encouragement to a crusading "holy war" mentality. The degrading experience of Western imperialism alone should be enough to discourage efforts to force other peoples to be free. Such efforts are likely to be neither just nor successful.

A second temptation, related to the first, may be to define "democracy" too narrowly and ethnocentrically, equating it too readily with all the particular norms and institutions of the Western parliamentary tradition. True, the norms and institutions of democracy as westerners know it have provided powerful restraints on absolutism. They can be treated as *an* effective model for others to adopt. But it is better for those norms and institutions to be adapted to the conditions of other histories and cultures than to be adopted as copied from a Western template. If the goal is one of a world where all peoples experience a high degree of self-determination and consent of the governed, norms and institutions that flow out of other peoples' histories may succeed on soils where particular Western forms would not. For this line of thinking we should relax the rigorous (and consciously ethnocentric) operational definition of democracy used early in this chapter.

It is inexcusably ethnocentric to imagine that other peoples are inherently incapable of autonomy and self-government, to declare them unsuited for democracy. Although it is myopic to overlook or idealize the ways in which many Third World governments, for instance, oppress their own peoples, it is equally ethnocentric to imagine that their ways of insuring autonomy and self-determination will be exactly like ours or to require the fully panoply of Western forms. In terms of the vision here, what is important is to support democratic governments where they exist and to recognize and reinforce a worldwide movement toward greater popular control over governments, rather than to specify the end-point in detail for each case.

HUMAN RIGHTS AND INFORMATION FLOWS

Whatever the faults of Western liberal (bourgeois) democracy, a world of spreading democratic ideology and practice offers some significant possibilities for spreading peace. Those possibilities can be enhanced by attention to implementing a broad definition of human rights and institutionalizing greater information flows. Human rights and information constitute elements both of greater global democratization and of direct and indirect contributions to international peace. In a world of imperfect democratization, such measures can help reduce those imperfections and compensate for some of them in the avoidance of war.

Commitment to Human Rights

Recent US governments have tended, in different ways, to emphasize a commitment to human rights. In the Carter Administration this began with an emphasis on political rights and civil liberties throughout the world; US standards were applied both to communist countries and to Third World States. Those governments found wanting did not appreciate the criticism. US attention to human rights in the Soviet Union reflected and perhaps hastened the decline of detente; despite some successes in the Third World, US pressures often angered allies thought to be strategically important, and the pressures were lessened. During the early years of the Reagan administration, official policy on human rights seemed to be turned most critically toward the Soviet Union and its allies, with abuses by US allies typically overlooked, tolerated, or even abetted. US allies were said to be merely authoritarian States, not totalitarian ones. The frequent ineffectiveness or hypocrisy of US human rights policy has given the whole concept a bad name to some otherwise sympathetic and liberal-minded people. But the forces strengthening human rights can at least be assisted by low-key persuasion and good example.

External Pressures to Promote Human Rights

Efforts to promote human rights internationally have not been uniformly ineffective or hypocritical. Third World governments sometimes do relax the worst of their oppression in response to external pressures, whether those pressures come from governments, international organizations, or private transnational organizations like Amnesty International and Americas Watch. External pressures can contribute to the legitimacy of internal opposition. Some of the rhetoric and liberalizing action of Gorbachev owes a great debt to the power and attractiveness of Western concepts of human rights. Western efforts to reiterate those concepts and their implications — for Eastern Europe as well as for the Soviet Union itself — can hardly be abandoned. An image of the Soviet government as willing to grant a fairly high degree of autonomy to its own citizens but not to its neighbors would hardly fit the image of a State with a "liberal," live-and-let-live policy essential to the basis of international peace being discussed in these pages.

Yet political concessions in the form of domestic human rights policies cannot be *demanded* of another greater power. The principle of non-interference in the internal governance of other States (statist and positivist norms in international law), dating from the end of the Thirty Years' War, does help defuse one major source of interstate conflict and cannot lightly be cast aside. Hectoring or badgering the leaders of another great power is likely to poison political relations and exacerbate other conflicts. Linkage of human rights concessions to important arms control measures, for example, is likely to hobble efforts to reduce real dangers of inadvertent conflict escalation. The failure to reach human rights goals should not become a reason to forego arms control agreements nor, worse, used as an excuse to prevent arms control agreements.

International Dialogue on Human Rights

International discussions on human rights are properly a dialogue, wherein the normatively persuasive elements are not solely those of Western advocates. A broad conception of human rights most certainly requires great emphasis on the kind of political rights stressed in US statements. Movement toward a more democratic world requires continued repetition of that message. It also requires a recognition of the legitimacy of some of the rights stressed by others: economic rights — to employment, housing, and some basic standard of material life.[33] Justice demands political liberty, and it also demands a level of economic decency. Political and social peace within democratic countries has been bought in part by this recognition; severe dismantlement of the welfare state would inflame class and ethnic conflict, and most elected political leaders know it. Internationally, recognition of the multifaceted nature of human rights is essential if the dialogue is to be one of mutual comprehension and persuasion rather than a dialogue of the deaf. This is a way in which political rights, economic rights, and international peace are bound inextricably together.

The Need for Economic Prosperity and Justice

Increasing worldwide adherence to democratic political norms and practices cannot alone bear all the weight of sustaining peace. Greater prosperity and economic justice, especially in the Third World, must also bear a major part. This conviction has often been expressed;[34] cynics often dismiss it. But it is unlikely to be merely a coincidence that, as noted earlier, the industrial democracies are rich as well as democratic. The distribution of material rewards within them, although hardly ideal, is nevertheless far more egalitarian than that within many Third World countries or between First and Third World peoples. A relatively just distribution does affect the cost-benefit analysis of those who would drastically alter it by violence; both rich and poor know they could lose badly. Some such calculation, including but not limited to the normative demands of justice, must apply to cement peace between nations. The broader human rights dialogue, incorporating political, cultural, and economic rights, constitutes a key element of global democratization where the domestic institutions of democracy are imperfect.

Practices and Institutions

Another aspect of a stable international peace—reinforcing but not fully contained in concepts of political democracy and human rights—concerns practices and institutions for international communication and cooperation. This has several elements.

International Trade and Cooperation

One element is *economic*: a freer flow of goods and services between communist and capitalist countries, especially the Soviet

Union. Henry Kissinger's detente policy envisaged such a network
of interdependence, giving the Soviet Union a greater material stake
in peaceful relations with the capitalist world, and increased Soviet
interest in Western products and markets makes the vision all the
more plausible. The vision is consistent with traditional liberal
prescriptions for trade and international cooperation.[35] Although it
is not a sufficient condition for peace, and possibly not even a necessary one, it certainly can make an important contribution.

Free Flow of Information

Economic exchange is also a medium and an occasion for the
exchange of *information*. Facilitation of a freer flow of information
is a second major element. Without a free flow of information
outward there can be no confidence that democratic practices are
really being followed within the other country, and sharp restrictions
on the flow of information into one's own country are incompatible
with the full democratic competition of ideas inside it. Cultural
exchanges and freer travel across State boundaries can help ease
misunderstandings of the other nation's reasoning, goals, and intent.
Across the spectrum from academic game theory to concrete social
experience we know that the prospects for cooperation are much
enhanced if the relevant actors can communicate their preferences
and actions clearly. This too is not a sufficient condition, and it is
easy to trivialize or ridicule the idea by imagining that communication alone can solve international problems. But without the
dependable exchange of information, meaningful cooperation is
virtually impossible in a world of complex problems and complex
national governing systems.[36]

Organizations as Centers of Shared Interests

It is in this sense that *institutions* — especially what Robert
Keohane calls "information-rich" institutions[37] — are valuable as a
means to discover and help achieve shared and complementary
interests. Global organizations such as UN agencies are important
purveyors of relevant information. Regional organizations, especially
among culturally similar countries, may be much less important as
instruments of coercion or enforcement than as a means of spotlight-
ing major human rights violations and upholding the moral force of
higher norms. The European Commission on Human Rights and the
European Court of Human Rights have done this effectively; the

Inter-American Commission on Human Rights and the Inter-American Court of Human Rights to a lesser degree. Transnational and populist legal norms serve to counter statist ones, and principles of democratic rights become incorporated, often through treaties, into international law and thereby into other States' domestic law.[38]

Communication Can Enhance Trust

The element of information exchange relates directly to *security* issues. Arms control and disarmament agreements require confidence that compliance with the agreements can be verified. Arrangements for insuring verification must be established on a long-term, reliable basis. Without it the agreements are continually hostage both to real fears that the agreements are being violated and to the pernicious charges by those who are opposed to the agreements whether or not they are being violated. Authoritarian governments can more easily, if they wish, pursue long-term strategies of aggressive expansion than can pluralistic democracies with many power centers and public discussion. "Democratic governments can also have their military buildups, of course, but cannot mask them because a public atmosphere of fear or hostility will have to be created to justify the sacrifices; they can threaten other countries, but only after their action has been justified in the open. . . ."[39] Liberalization of the Soviet Union will allow its external partners/adversaries to feel less apprehensive and to feel more confident that they will have early warning of any newly aggressive policy.

A dense, informal network of information exchange that extends across a wide range of issues and is beyond the control of any government will help, as will some formal institutions for information-sharing. Just as substantial freedom of information is essential to democratic processes within a country, it is essential to peaceful collaboration between autonomous, self-determining peoples organized as nation-states.

Certain specific kinds of multilateral institutions can be important in controlling crises. One possibility is creation of crisis management centers, along the lines agreed to for the United States and the Soviet Union but extended to include other nuclear powers whose actions might precipitate a crisis. Another possibility is strengthening the information and communications base — now sadly inadequate — of the UN, and especially of the Secretary General so that in some future event like the Cuban Missile Crisis he might be available, in timely and informed fashion, as a valuable mediator. Yet another possibility is the operation of observation satellites by third parties (other countries or international organizations) to monitor military activities and arms control compliance by a variety of electronic means.[40] As long as nuclear weapons exist, even in a world of substantial political liberalization, reliable means of information exchange will be essential.

THE COMING TEST?

Democracies, as well as other political systems, do have their dark side of externalizing popular frustrations, and some degree of xenophobia is virtually universal. The United States and the Soviet Union, as multi-ethnic societies, are perhaps especially prone to defining patriotism in terms of loyalty more to the political system than to a set of cultural principles. Democracy and socialism thus become defining principles for identifying friends and foes. A shift away from seeing the other as the enemy will not come easily,[41] though it may be assisted by change within the Soviet system, which blurs the existing differences of political practice.

The winds of democracy are blowing in the world, even if in gusts of variable strength and direction. Recently they have become especially evident in the communist countries and dramatically so with *glasnost* and *perestroika* in the Soviet Union. The current depth and long-term prospects for this movement are highly uncertain. Here too, the twin dangers of wishful thinking and willful ignorance about these events are unavoidably present. The Soviet Union may not soon become a political democracy as people in the West understand the term. The bureaucratic, cultural, and historical constraints are powerful.

But democratization as a process is (as of the early 1990s) surely occurring. There are greater freedoms of expression and dissent within the Soviet Union and greater degrees of competition for political control. There is greater openness across the Soviet Union's international borders for the transmission of ideas and information into as well as out of the country. Prospects for greater trade and cultural exchange can help solidify this openness. Mikhail Gorbachev is explicitly asking his country to adopt some Western norms, as desirable in themselves and as legitimating economic and political modernization. This is an exhilarating and uncertain process, perhaps subject to some reversal but not easily controllable by any leader or group. With the ideological norms destroyed, practice can never be the same.

An article of faith of the dominant ideologies in both the United States and the Soviet Union has always been that neither has any fundamental quarrel with the people of the other country. US differences allegedly have been not with the Soviet people, but with the atheistic communist elites that repress the people; alternatively, Soviet differences have been not with the US people but with the greedy capitalist elites who exploit them. Insofar as the Soviet system of government operated under principles so manifestly different from those of Western democracy, the US claim had a *prima facie* validity.[42] And insofar as the Soviet leadership explicitly rejected the legitimacy of Western democracy as representing people's interests, their claim also seemed valid to their people.

Attitudes of the US public toward China have shifted greatly over the past 20 years, from an evaluation that was even more negative than the one held toward the Soviet Union to one that is now, on balance, mildly positive. This shift, particularly until 1989, coincided with the moderate relaxation of internal controls within China. A similar shift in attitudes – so far weaker – has begun in US attitudes toward the Soviet Union, again coincident with and very probably assisted by the limited changes that have occurred in Soviet internal politics.[43] As Soviet ideology and practice begins to shift, the distinction between ruling elites and their people loses some of its force. If both sides see each other as in some sense truly reflecting the rule of law and the consent of the governed, the transformation of international relations begins.

CONCLUSION

Realist theories about the inherently antagonistic structure of international relations have never been tested in a world where all the major States were governed more or less democratically. Thus we never have had a proper test of some realist propositions against liberal idealist ones.[44] Perhaps we are about to see one. Even if liberal idealist theories are correct, it is not clear what threshold of democratic norms and practices must be crossed to achieve peace or whether it is merely a matter of greater *degree* of democratization bringing a greater *likelihood* of peace between States. It also is not clear what ancillary conditions must be met or whether the necessary threshold can in fact be reached by anything that can evolve from the cold war system. A hint, however, is that among characteristics deemed absolutely necessary for cooperative relations in mid-1988, only 29 percent of US citizens insisted that the other country be "a democracy," but 53 percent required that it be one where "citizens enjoy basic human rights."[45]

Perhaps two or more great powers can exist in the same international system where, governed by mutual self-interest and some sense of broader interest but not ruled by any superordinate authority, they can build a system for war avoidance that does not depend primarily on nuclear deterrence and military threat. If so, the results will be of far more than academic interest. They may provide the most reliable and most readily available system of alternative security. Nuclear weapons might continue to exist, but crises would occur less often and, when they did, would carry less weighty ideological baggage. From the perspective of the cold warriors on both sides of what was called the Iron Curtain, the really subversive nature of *glasnost* may be that it will make the Soviet Union no longer eligible for "the presumption of enmity."

NOTES

An earlier version of this chapter, which was prepared originally for this volume, was published in Volume 26 of *World Futures* (1990). Copyright © 1990 by Gordon and Breach. It is reprinted here, with changes, with the permission of *World Futures*. For comments on early drafts, I am indebted to Robert Dahl, Lloyd Dumas, Raymond Gastil, David Lumsdaine, Bradford Westerfield, and Burns Weston—who of course bear no responsibility for the result. The subtitle is taken from the epigram of Seymour Martin Lipset, *Political Man* (Garden City, NY: Doubleday, 1960). When I first read Lipset's aphorism I found it highly implausible.

1.   *See* Richard J. Bernstein, *Beyond Objectivism and Relativism* (Philadelphia: University of Pennsylvania Press, 1983), and Thomas J. Haskell, "The Curious Persistence of Rights Talk in the 'Age of Interpretation,'" *American Historical Review* 74, 3 (December 1987): 984-1012.

2. *See* Karl W. Deutsch *et al.*, *Political Community and the North Atlantic Area* (Princeton, NJ: Princeton University Press, 1957), characterizing this situation as a "security community"; and Kenneth E. Boulding, *Stable Peace* (Austin, TX: University of Texas Press, 1979), characterizing it as "stable peace." In duration and expectation it differs from the simple absence of war that may prevail between some other States, including non-democratic ones in the Third World.

3. John Mueller, *Retreat from Doomsday: The Obsolescence of Major War* (New York: W. H. Freeman, 1989, 3rd edition): ch. 14.

4. These attempted explanations are considered at greater length in Bruce

Russett and Harvey Starr, *World Politics: The Menu for Choice* (New York: Basic Books, 1989, 3rd edition): ch. 14.

5. Bruce Russett, *International Regions and the International System* (Chicago: Rand McNally, 1968): ch. 12; and David Wilkinson, *Deadly Quarrels: Lewis F. Richardson and the Statistical Study of War* (Berkeley, CA: University of California Press, 1980): ch. 9.

6. *See* Peter J. Fliess, *Thucydides and the Politics of Bipolarity* (Nashville, TN: Parthenon Press, for Louisiana State University Press, 1966): 131. In classical and medieval times the State, even in a democracy, was seen as actively shaping society rather than as some impartial arbiter. Hence such States, in addition to sharply restricting the franchise, lacked the modern concept of citizens' natural rights. Thus their behavior provides but an imperfect test of the theory here. *See* Harvey Mansfield, Jr., "On the Impersonality of the Modern State," *American Political Science Review* 77, 4 (1983): 849-57.

7. *See* Melvin Small and J. David Singer, "The War-Proneness of Democratic Regimes," *Jerusalem Journal of International Relations* 1, 1 (1976).

8. *See* Robert A. Dahl, *Polyarchy: Participation and Opposition* (New Haven, CT: Yale University Press, 1971).

9. *See principally* Steve Chan, "Mirror, Mirror, on the Wall . . . Are the Freer Countries More Pacific?," *Journal of Conflict Resolution* 28, 4 (1984): 617-48; Michael Doyle, "Liberalism and World Politics," *American Political Science Review* 80, 4 (1986): 1151-61; Zeev Maoz and Nasrin Abdolali, "Regime Types and International Conflict, 1816-1976," *Journal of Conflict Resolution* 29, 1 (1989): 3-35; R. J. Rummel, "Libertarian Propositions on Violence within and between Nations: A Test Against Published Research Results," *Journal of Conflict Resolution* 29, 3 (1985): 419-55, and "Libertarianism and International Violence," *Journal of Conflict Resolution* 27, 1 (1983), 27-71; Small and Singer, "War-Proneness of Democratic Regimes": 50-69; Peter Wallensteen, *Structure and War: On International Relations, 1920-1968* (Stockholm: Raben and Sjogren, 1973); and Erich Weede, "Democracy and War Involvement," *Journal of Conflict Resolution* 28, 4 (1984): 649-64.

10. *See* note 9. Of the authors cited, Rummel is the only dissenter on this point, chiefly on the basis of a study of 1976-80, a period that omits, among others, Vietnam and most post-colonial wars.

11. For a good introduction to Gramscian ideas, see T. J. Jackson Lear, "The Concept of Cultural Hegemony: Problems and Possibilities," *American Historical Review* 90, 3 (June 1985): 567-94. Of course, if the governments are not broadly representative—or if one of them is not—despite cultural belief to the contrary, the possibility of unreconcilable conflicts of interest between them is increased.

12. *See* Kenneth Waltz, *Theory of International Politics* (Reading, MA: Addison Wesley, 1979).

13. Other countries are *liked* also to the degree they are democratic. In 1986 a national survey asked Americans, in a commonly used question, to rate the warmth or coolness of their feelings toward other countries. Of the top

ten countries so rated, all but two were democracies and the exceptions only partially non-democratic (Mexico, fifth, and Brazil, tenth). Of the lower thirteen, all but two were clearly not democracies (exceptions being India, seventeenth and Nigeria, nineteenth). *See* John Reilly, *American Public Opinion and U.S. Foreign Policy 1987* (Chicago: Chicago Council on Foreign Relations, 1987).

14. *See,* for example, *Gallup Reports*, Nos. 83 and 96 (July 1972 and June 1973).

15. Bruce Russett and Thomas W. Graham, "Public Opinion and National Security Policy: Relationships and Impacts," in Manus Midlarsky (ed.), *Handbook of War Studies* (Winchester, MA: Unwin Hyman, 1989).

16. Doyle "Liberalism and World Politics": 1161.

17. There have also been cases of covert intervention (rather than overt attack) against some radical but elected Third World governments (*e.g.*, Guatemala, Chile), justified by a cold war ideology and public belief that the government in question was allying itself with the major non-democratic adversary.

18. Cynthia Eagle Russett, *Sexual Science: The Victorian Construction of Womanhood* (Cambridge, MA: Harvard University Press, 1989); R. J. Vincent, "Racial Equality," in Hedley Bull and Adam Watson (eds.), *The Expansion of International Society* (Oxford: Clarendon, 1984); Hannah Arendt, *Imperialism* (Part 2 of *The Origins of Totalitarianism*) New York: Harcourt Brace, 1952).

19. Aid to the Nicaraguan Contras never was really popular in the United States, not just because of *realpolitik* fears of "another Vietnam" but because of a general perception that the Contras were no more legitimate or representative of their people than were the Sandinistas.

20. Helmut Norpoth, "Guns and Butter and Government Popularity in Britain," *American Political Science Review* 81, 3 (1987): 949-60.

21. Charles W. Ostrom and Brian L. Job, "The President and the Political Use of Force," *American Political Science Review* 79, 2 (1986): 541-66; Bruce Russett, "Economic Decline, Electoral Pressure, and the Initiation of Interstate Conflict," in Charles Gochman and Alan Ned Sabrosky (eds.), *Prisoner of War? Nation-States in the Modern Era* (Lexington, MA: D. C. Heath, 1990).

22. Theodore Lowi, *The Personal President* (Ithaca, NY: Cornell University Press, 1985): 20.

23. *See ibid;*; *see also* John Mueller, *War, Presidents, and Public Opinion* (New York: Wiley, 1973).

24. Timothy Y. C. Cotton, "War and American Democracy: Electoral Costs of the Last Five Wars," *Journal of Conflict Resolution* 30, 3 (1986): 616-35.

25. Guillermo O'Donnell, "Challenges to Democratization in Brazil," *World Policy Journal* 5, 2 (1988): 281-300, for example, characterizes Brazil as in danger of becoming a "democradura;" *i.e.*, "a civilian government controlled by military and authoritarian elements."

26. Raymond Gastil, *Freedom in the World: Political Rights and Civil*

Liberties, 1988-1989 (New York: Freedom House, 1989 and previous annual editions).

27. For explication of first, second, *and* third generation rights, see Burns H. Weston, "Human Rights," in Richard Pierre Claude and Burns H. Weston (eds. and contribs.), *Human Rights in the World Community: Issues and Action* (Philadelphia: University of Pennsylvania Press, 1989): 16-20.

28. In the most recent edition, the author abandons his earlier three-part rating and relies solely on the fourteen point scale. For our purposes, however, it is appropriate to retain his previous equation of scores of 2-5 with "free," 6-11 with "partly free," and 12-14 with "not free." One can reasonably argue with elements of the author's ranking system, but relative to this chapter most of these arguments are not relevant. It has been applied quite consistently over the years, except that the author suggests he probably should not have coded several Latin American states (Colombia, El Salvador, and Guatemala) as "free" in 1973. If so, however, that inconsistency *understates* the substantive shift toward democracy indicated by Table 4.2. His judgments also agree well with independent judgments by Michael Coppedge and Wolfgang Reinecke, "A Scale of Polyarchy," in Gastil's *Freedom in the world, 1988-1989*; Dahl, *Polyarchy*: Appendix B; and authors cited in note 9.

29. Robert A. Dahl, "Democracy and Human Rights Under Different Conditions of Development," paper for the Nobel Symposium on Human Rights (Oslo, Norway, June 1988).

30. George Modelski, "Is America's Decline Inevitable?" *Uhlenbeck-Lecture VI* (Wassenaar: Netherlands Institute for Advanced Study, 1988).

31. Eric Hanson, *The Catholic Church in World Politics* (Princeton, NJ: Princeton University Press, 1987).

32. Arthur A. Stein, *The Nation at War* (Baltimore: Johns Hopkins University Press, 1980).

33. *See* Charles R. Beitz, *Political Theory and International Relations* (Princeton, NJ: Princeton University Press, 1979); Samuel S. Kim, *The Quest for a Just World Order* (Boulder, CO: Westview, 1984). *See also* Claude and Weston, *Human Rights in the World Community*.

34. For example, Willi Brandt, *North-South: A Program for Survival* (Cambridge, MA: MIT Press, 1980), and Henry Shue, *Basic Rights: Subsistence, Affluence, and U.S. Foreign Policy* (Princeton, NJ: Princeton University Press, 1980).

35. Richard Rosecrance, *The Rise of the Trading State* (New York: Basic Books, 1986).

36. *See* Bruce Russett, *Community and Contention: Britain and America in the Twentieth Century* (Cambridge, MA: MIT Press, 1963).

37. Robert O. Keohane, *After Hegemony: Cooperation and Discord in the World Political Economy* (Princeton, NJ: Princeton University Press, 1984): 247.

38. *See* Richard Falk, *Human Rights and State Sovereignty* (New York: Holmes & Meier, 1981): ch. 3. *See also* Myres S. McDougal, Harold D. Lasswell, and Lung-chu Chen, *Human Rights and World Public Order* (New

Haven, CT: Yale University Press, 1980), and Francis Anthony Boyle, *World Politics and International Law* (Durham, NC: Duke University Press, 1985).

39. Edward Luttwak, *Strategy: The Logic of War and Peace* (Cambridge, MA: Belknap, 1987): 235.

40. Thomas E. Boudreau, *The Secretary General and Satellite Diplomacy* (New York: Council on Religion and International Affairs, 1984); Ann Florini, "The Opening Skies: Third-Party Imaging Satellites and U.S. Security," *International Security*, 13, 2 (1988): 91-123.

41. R. Paul Shaw and Yuwa A. Wong, *Genetic Seeds of Warfare: Evolution, Nationalism, and Patriotism* (Boston: Unwin Hyman, 1989).

42. Almost 90 percent of Americans believe that the Russian people are not as hostile to the United States as are their leaders and that the Russians could be our friends if the attitude of their leaders was different. Daniel Yankelovich and Sidney Harman, *Starting with the People* (Boston: Houghton Mifflin, 1988): 64, citing a December 1983 survey.

43. Bruce Russett, "Democracy, Public Opinion, and Nuclear Weapons," in Philip Tetlock, Jo L. Husbands, Robert Jervis, Paul Stern, and Charles Tilly (eds.), *Behavior, Society, and Nuclear War* (New York: Oxford University Press, 1989).

44. Neither realist nor liberal idealist theories are fully adequate, but the dominance of realist thinking in contemporary academic as well as government circles has tended to diminish attention to realism's analytical and empiricial weaknesses. *See* Joseph R. Nye, Jr., "Neorealism and Neoliberalism," *World Politics* 40, 2 (1988): 235-51, and John Vasquez, "The Steps to War: Toward a Scientific Explanation of Correlates of War Findings," *World Politics* 40, 1 (1988): 108-45. Note that the theoretical perspective of my chapter attends neither to the international system level of analysis nor to the individual nation-state but rather to the nature of *relations* between two States. For the distinction, see Russett and Starr, *World Politics*: ch. 1.

45. Daniel Yankelovich and Richard Smoke, "America's 'New Thinking,'" *Foreign Affairs* 67, 1 (1988): 16.

5

Economics and Alternative Security: Toward a Peacekeeping International Economy

Lloyd J. Dumas

You cannot organize civilization around the core of militarism and at the same time expect reason to control human destiny.

— Franklin Delano Roosevelt, 1938

At least one important school of anthropological thought assures us that we humans are not inherently violent.[1] But even if we were, this would not by itself guarantee that war must be a continuing part of the human enterprise. War is a social institution; it is not merely the eruption of interpersonal violence on a larger scale. It is coordinated, organized mass violence—violence that has become less and less personal as the years have gone by. And as it has become less and less personal and more and more guided by calculated bureaucracies and technologies, so also has it become more destructive.

The military is a social institution, too. Social institutions are artificial constructs created to serve human purposes, and the purpose of the military—its only legitimate purpose—is to provide security by ensuring peace and protecting a way of life. Most of us believe that our own military does just that, and for this reason we have lavished resources and attention on our military forces in a spectacularly successful effort to make them ever stronger and more destructive.

Paradoxically, however, our very success has propelled us into a world in which our peace and security—even our survival—is problematic. Put simply, I believe that military force and the threat of military force, as a primary means of ensuring the peace and security all of us so deeply desire, is an idea whose time has gone. This is particularly so where nuclear weapons are involved. War is too destructive and the preparation for war too expensive for human society to continue to rely on the brute force of large militaries to secure its safety. Since the dawn of the nuclear age, the handwriting has been on the wall: Either we put an end to militarism and war or militarism and war will most assuredly put an end to us. Those who continue to rely on nuclear deterrence and other threats of mass violence as if this were not so are utopianists—dreamers—in the worst sense. They live in a fantasy world.

Because war-threatening military force is a social institution that no longer serves its purpose, it is the essence of realism to look for other, more effective social arrangements. There is nothing more realistic or pragmatic than discarding that which no longer works in favor of that which does. To be sure, it is easy to be cynical about the prospect of finding a viable alternative to nuclear deterrence or to war itself; nothing seems so immutable as the status quo. And yet the one constant in human history is change. Once slavery was a generally accepted social institution, with its own set of vested interests, deeply embedded in the status quo. Today, there is not a single nation on earth in which it is legal for one human being to own another. The very idea seems absurd. So can it be with militarism, nuclear deterrence, and even war itself.

Of course, slavery did not disappear on its own. It ended for many reasons: changing attitudes, changing technology, and changing political and economic conditions. But it ended also because there were people who believed that human beings could do better, and they were willing to combine their vision of a world without slavery with the hard work of building a practical path to their dream.

Like my co-authors in this volume, I believe human beings can do better when it comes to ensuring our national and international security. The "Machines of Armageddon"[2] are not and need not be our sole recourse. In this chapter, I probe the ways in which economics can help in the search for more effective, alternative approaches to national and international security.

ECONOMIC RELATIONSHIPS AND CONFLICT

There has long been disagreement as to the connection between economic relationships and violent conflict. The debate goes back to the Eighteenth Century arguments between the mercantilists and Adam Smith and carries through the arguments between Marxists and economic liberals today. Some (*e.g.*, mercantilists) have argued that threats of violent coercion and the use of force are effective means of pursuing economic benefit and national power. Others (*e.g.*, Marxists) have argued that the forceful expansion of economic activity abroad, to the extent of armed conquest if necessary, is critical to relieving social pressures at home (created by unjust economic systems) that otherwise would erupt into full-blown economic and political crises. Thus, according to these schools of thought, economic factors—control of valuable natural resources, access to markets, or simply the desire to take valuable goods and services from others—have been major driving forces in generating conflict and war.[3]

Beyond this, economic relationships can create conflicts that can and do grow to the point of explosion, erupting in civil, revolutionary, and international wars. These wars often shift economic relationships in such a manner as to give rise to other conflicts that sow the seeds of future wars. Those disadvantaged at the end of the last war sooner or later seek to gain the upper hand.

It also has been argued, however, that war and preparation for war tend to damage rather than benefit national economies, especially in the long run. Writing in 1776, Adam Smith contended that expenditures on such activities were economically unproductive.[4] I have argued that sustained high levels of military spending tend to undermine national economies by diverting critical economic resources needed for the maintenance and improvement of a nation's productive capacity.[5]

Furthermore, economic liberals generally believe that widened international economic relationships tend to create greater opportunities for mutual benefit through cooperation rather than confrontation. Thus international economic relations may well tend to reduce the likelihood of war.[6] In support of this position, Richard Rosecrance has written:

> While trading states try to improve their position and their own domestic allocation of resources, they do so within a context of accepted interdependence. . . . [T]hey prefer a situation which provides for specialization and division of labor among nations. One nation's

attempt to improve its own access to products and resources, therefore, does not conflict with another state's attempt to do the same. The incentive to wage war is absent in such a system for war disrupts trade. . . .[7]

A web of economic relationships binds the participants together. Their growing interdependence does not prevent conflicts, but it does create strong incentives to settle them amicably. Furthermore, ongoing economic relationships typically imply continuing personal interaction and contact, and this tends to break down prejudices, which lower the threshold of violent conflict. From this perspective, the strengthening and expansion of the global economy may be the single strongest force for world peace today.

Which of these positions is correct? Whether international economic relationships tend to raise or lower the probability of war depends on their nature, not just their extent. Broadly speaking, exploitative relationships – those in which the flow of benefit is overwhelmingly in one direction – tend to increase the number and severity of conflicts whereas mutually beneficial relationships – those in which the flow of benefit is more or less balanced – tend to reduce the likelihood and intensity of conflict.

The reason is straightforward. Exploitative relationships are inherently unfair. Even if the parties being exploited gain something from the relationship, the fact that the vast majority of benefit flows in the other direction is bound to create or aggravate antagonisms. In either case, the exploited stand to lose little and may even gain if the relationship is destroyed. They therefore may be inclined to raise the intensity of whatever economic or other conflicts might occur, even to the point of war.

Balanced relationships tend to have the opposite effect. All parties gain from the ongoing connection they have established. Out of pure self-interest, they do not want to see the relationship disrupted. Furthermore, balanced relationships permit, even foster, the full development of every party involved. As the economic condition of each participant improves, their producers become more productive and their potential as a market grows. Thus, their capacity to contribute to the relationship increases, both as a source of products and profits.

It is a great mistake to view international economic relationships as though there always is a fixed pie of benefit to be divided, the larger gain of one party coming only at the expense of another. Exploitative relationships do tend to have this character, but balanced relationships not only provide for mutual gain, they tend also to cause the pie of benefit to grow as time goes by. As a result, when relationships are balanced, the incentives are strong to settle conflicts amicably. And as they are successfully resolved time after time, the idea of allowing them to fester to the point of confrontation comes to seem more and more absurd. Under such circumstances, the thought of brandishing the threat of war slowly recedes, and war itself becomes ultimately unthinkable. The European Economic Community (EEC—the Common Market) is an interesting case in point.

By 1988, EEC membership included Belgium, Denmark, France, Germany, Greece, Ireland, Italy, Luxembourg, the Netherlands, Portugal, Spain, and the United Kingdom,[8] countries that have fought countless wars with each other over the centuries. Yet, today, if you were to ask their citizens what they thought the prospect of war between their homelands would be over the next fifty years, it would not be considered a serious question. Quite the contrary, trade, travel, and cooperation among their nations is growing; the bonds are strengthening as they move to eliminate all remaining trade restrictions by 1992 and contemplate establishing a common currency.

True, the military cooperation of many of these nations in NATO doubtless has played some part in diminishing the possibility of war among them. Yet membership in NATO—not in the EEC—did not stop Greece and Turkey from facing off against each other militarily during the Cyprus crisis of the mid-1970s. Furthermore, it is not uncommon for military alliances to dissolve quickly into serious military confrontation or war. The US-USSR shift from military allies to antagonists after World War II is a striking example. A comparable disintegration of the Common Market would be astonishing.

It is easy to see why someone being exploited in an unbalanced relationship would be better off if the relationship were to become more balanced. But the dominant party, over the long term, would gain from greater balance in most cases also. A dominating party in an exploitative relationship generally has to expend considerable effort to maintain control, and this effort is much more of a drain than is commonly supposed. When, however, a relationship is balanced, there is no need to expend extra effort to keep it going. The mutual flow of benefits binds the parties together. Thus, a balanced relationship is in effect a more efficient relationship; the benefits are achieved at much lower cost.

Of course, in any exploitative relationship the expense of maintaining control potentially can be reduced if the exploiters are able psychologically to manipulate the exploited to make them feel helpless or accepting—even deserving—of their subordinate position. This is the basis of what may be one of Karl Marx most important insights, the concept of "false consciousness"[9]—a concept that can be extended to an even more effective method of maintaining control, namely, getting people to identify with the very system that is exploiting them. If appeals to patriotism or to the grandeur of empire succeed in making people proud to be part of such a powerful institution, they not only may accept but actively support a system that is in fact exploiting them.

But even false consciousness does not work forever; and though the benefit to the dominator in an exploitative relationship *may* be greater than if the relationship were balanced, even after allowing for the cost of maintaining control, that tends to be true only in the short run. In the long run, continued exploitation breeds discontent and disaffection, the exploited come more and more to see the possibility and desirability of change, and the dominator in the exploitative relationship consequently suffers heightened insecurity and expense in maintaining control. As Adam Smith concluded more than two hundred years ago, after a lengthy discussion in the context of the British colonial empire:

> Under the present system of management . . ., Great Britain derives nothing but loss from the dominion which she assumes over her colonies.
> . . . Great Britain should voluntarily give up all authority over her colonies. . . . [She] would not only be immediately freed from the whole annual expense of the peace establishment of the colonies, but might settle with them such a treaty of commerce as would effectually secure to her a free trade, more advantageous to the great body of the people . . . than the monopoly which she at present enjoys.[10]

In sum, exploitation tends ultimately to become counterproductive.

But if it is true that exploitation is counterproductive in the long run even for the exploiter, why do we continue to believe that we are best off if we are dominant in a relationship? There are, I think, two main reasons.

The first and most obvious reason is myopia. Human beings can much more easily calculate and relate to the short-term effects of their actions than the long-run consequences. Even if a certain behavior is known to be counterproductive in the long run, it is very difficult to get people (or their nations) to stop if the short-term effects are thought desirable. Cigarette smoking and chronic overeating (or overdieting) come easily to mind as examples for individuals.

The second reason may well be yet more compelling. We come to know, virtually from birth, that for much of what we need or want in life we must depend on others. From infancy, our growth is largely a transition from total dependency to greater self-reliance. Yet even in adulthood, we remain intellectually, emotionally, and physically dependent on others for guidance, interaction, praise, and support, as well as for the material goods and services we want and need. This creates a degree of insecurity that comes from the fear that those on whom we must depend may not do what we want or need them to do. To some extent, we have learned to deal with our insecurity through a paradigm of domination. If we are in a position of control, we can manipulate anyone who will not go along voluntarily to give us what we want or need.

There is much in the human experience, as individuals and as nations, to support the idea that those who dominate often live at a higher material standard of living, at least in the short run. It is much harder to see the negative long-run consequences for the exploiters or that their dominance does not necessarily translate into a higher quality of life even in the short run—that is, when all of life's critical intangibles (such as love, self-respect, and feelings of fulfillment) are taken into account.

The paradigm of domination may never have been optimal in international affairs, but it has become increasingly dangerous, for the machines and institutions of nuclear war are the logical and inevitable creations of this paradigm. They are the ultimate coercive threat for the purpose of dominating others and the ultimate counterthreat to frustrate the attempts of others to dominate us. This system of threat and counterthreat has lead us inexorably closer to the brink of the war we all dread. It has become utterly self-defeating.

The threat or use of force is actually a much less effective way of getting others to do what you want them to do than taking a more positive approach—for example, by making friends or creating working partnerships.[11] It is an obvious lesson of everyday life that we fail to apply to the realm of international relations because our thinking has been so narrowed by the paradigm of domination. Suppose, for example, that some neighbors have been playing their stereo very loudly at night and that it really disturbs you. You could pound on their door and threaten to have them arrested if they do not stop the noise. It is possible they will stop, but it is even more likely that they will get angry and either make the music louder or find some other way of annoying you. Suppose instead you invited your neighbors over for coffee and took the time to try to establish a more friendly relationship. It is possible they would continue to play the music loudly, but it is much more likely that they would make the music softer. In fact, they are likely to turn the volume down as soon as they become aware that the music is bothering you. And to the extent that you do become friends, they generally will try to avoid creating disturbances in the future.

Of course, not everyone is reasonable. It is not always possible to build more friendly relations. There are circumstances in which coercion may be required. But the establishment of friendly relations is much more likely to be effective than threatening to use or actually using force. The use of military deterrence (nuclear or otherwise) is inherently threat-based, coercive, and unbalanced. It is an inferior way of creating security or of achieving other national objectives. If it is possible to strengthen positive incentives for the peaceful resolution of conflicts, that is very likely to be a less costly and more effective path to increased security and greater economic well-being.

When is it possible to replace more distant or hostile relationships with friendlier ones and how might that be done? Across the spectrum of relationships from the personal to the political, these are not easy questions to answer. Clearly, willingness and commitment to undertake initiatives aimed at breaking down barriers play a key role.[12] But factors internal to the parties may also be critical in increasing receptiveness to friendlier ties. Both of these considerations were obviously at work in dramatically improving relationships between the United States and the People's Republic of China (PRC) in the 1970s and between the United States and the Soviet Union in the late 1980s. Political and economic changes within the PRC, in part resulting from internal economic problems, greatly increased the willingness of the Chinese to respond to initiatives from the United States. Similar pressures and changes in the USSR pushed the Soviets to make a startling series of initiatives to establish friendlier relations with the United States, even in the face of an extremely hostile US administration — and these initiatives clearly succeeded.

STRATEGIES FOR A PEACEKEEPING ECONOMY

In his intriguing book *Stable Peace*, Kenneth Boulding sets forth what he calls the "chalk theory" of war and peace:[13] a piece of chalk breaks when the strain applied to it is greater than its strength, *i.e.*, its ability to resist that strain. Similarly, war breaks out when the strain applied to an international system exceeds the ability of that system to withstand strain. The establishment of "stable peace" thus requires that strains be reduced, that strength be increased, or both.

In maximizing the peacekeeping potential of the international economic system, it seems most fruitful to look for a combination of strain-reducing and strength-enhancing strategies. The four basic strategies that follow are, I believe, key to moving toward a peacekeeping international economy.

Strategy I: Balance Independence and Interdependence

Increasing independence tends to reduce strain because it reduces vulnerability to externally generated disruptions of the nation's economy. Vulnerability tends to create insecurity for a number of reasons. There is the fear that opponents will purposely exploit vulnerability to gain the upper hand in a conflict (*e.g.*, by cutting off the flow of key goods or resources) or that others will unintentionally harm you while pursuing their own objectives (*e.g.*, by inventing substitutes for the exports on which your economy depends). These fears may be exaggerated or they may be justified, but in either case the insecurity they generate typically produces defensive, belligerent behavior. A greater degree of independence helps to avoid these problems.

Increasing mutual interdependence tends to increase strength because it increases the incentive to avoid disruption. The web of interdependence created by balanced relationships yields a flow of mutual benefit that will be reduced or eliminated if the conflicts that arise are permitted to get out of control. This cannot help but encourage the peaceful resolution of conflict, increasing the resistance of the international system to strain.

How is it possible to create an economic system whose actors are both independent and interdependent? One approach is to reduce dependence in those areas of interaction in which vulnerability is most frightening while increasing it in other areas where there is a real potential for mutual gain. There is no doubt that dependence on outside sources of supply is more troubling the more critical the good or service in question. A nation is more exposed and vulnerable if it must rely on foreign suppliers for critical elements of its food supply than if it is a large net importer of television programs.

Now it might be argued that a nation that is dependent on an outside supplier of critical goods would be even less likely to go to war with that outside supplier than if it were independent. But unless the critical goods dependence is mutual, the nation *supplying* the critical goods has no comparable incentive to avoid war or other coercion against its customer. And, as argued earlier, it might even feel encouraged to behave coercively because of the leverage the criticality of the goods provides. The nations of the Organization of Petroleum Exporting Countries (OPEC), acting jointly, gained considerable economic and political benefit by using their critical goods leverage in the 1970s. Of course, the possibility that critical goods suppliers might try to use their leverage in this way gives strong incentives to the dependent nations to try to keep the supplying nations under control. Over time, a range of coercive policies from mild to severe have been used for this purpose, from manipulating corrupt officials to forced colonization.

Beyond this, one nation's dependence on another for critical goods might encourage a third party to behave more aggressively in dealing with the dependent nation when conflicts erupt. Why? Because the third party may be confident that it can disrupt the flow of critical goods and thereby quickly gain the upper hand, particularly if the critical goods supplier is itself a weak nation located far from its customer. This was part of the Japanese strategy in World War II. For all these reasons, the goal of independence in essential goods seems worthwhile.

Each nation's attempt to balance its independence and interdependence flows from its attempt to balance the political gains of reduced vulnerability against the economic gains of trade. It is not necessary to assume that nations act out of an unselfish desire to do their part in maximizing the economic well-being of the world as a whole. The pursuit of more narrowly defined self-interest is sufficient.

Which goods and services are most critical to a nation's economic well-being depends in large part on the nature of its economy and the level of its development. Reliable supplies of steel are far more critical to an industrialized nation than to one that is mainly agricultural; reliable supplies of tractors and harvesting machines are more critical to a highly developed agricultural nation than to either a nation that is largely industrial or one that is agricultural but much less developed. There are, however, some categories of goods that may be broadly considered critical.

Foods that have few substitutes and that play a central role in the diet of the population are clearly in the category of critical products. Thus it is desirable that each nation be capable of meeting the basic nutritional requirements of its own population. Similarly, potable water is critical to health and well-being. This fact not only implies that each nation should be able to provide minimum water requirements without depending on foreign suppliers but also that nations should be capable of ensuring the purity of that water. Energy supplies are critical to any economy, though they tend to be more completely integrated into those economies that are more highly developed. And every economy requires continuing supplies of certain key raw materials. Thus, independence in terms of energy and critical raw materials is desirable as well.

It is not necessary for each nation to be *totally* self-sufficient in critical goods. To avoid the insecurity inherent in dependence on an external lifeline of critical goods, a nation need only be able to supply the minimum amount of those goods it requires from internal sources during situations of prolonged disruption of trade. For example, a nation need not be able to produce locally its entire supply of food; it need only be able to supply sufficient food to meet the basic nutritional requirements of its population if and when it is cut off from foreign sources.

Nations unable to achieve independent domestic production of even minimum quantities of critical goods may still be able to achieve a degree of independence in domestic supply. When the critical goods are storable, it may be feasible to establish national stockpiles that provide a significant short-term buffer. That would at least buy time for a more reasoned response in the event of a sudden and arbitrary disruption of supply. If the buffer were sufficiently large, enough time would be available to pursue a number of avenues for the non-violent resolution of the conflict that brought about the disruption. It also would allow time for arranging alternative sources of supply. A large buffer might even serve as a deterrent to punitive disruption because the nation that contemplates interrupting the supply would know in advance that its action would be relatively ineffective. Similar results could be achieved also by developing a standby domestic production capability for critical goods.

Switzerland has adopted a combined standby domestic production capability and "strategic stockpiles" approach as part of its comprehensive national defense strategy. As Dietrich Fischer has written:

> During peacetime, Switzerland imports nearly 50% of its food consumption. In case of a cutoff of imports, food would be immediately rationed, requiring about a one-third reduction in daily caloric intake. Meat consumption would be drastically reduced to arrive at a more efficient calorie conversion factor through a more vegetarian diet. . . . Grassland would be converted gradually, in three yearly phases, into cropland. In the meantime, the food deficit would be bridged with reserves of non-perishable food, which are constantly being renewed in peacetime. Similar programs exist for fuel, certain minerals, and other vital commodities.[14]

On the other hand, interdependence should be maximized with respect to non-critical items. There are many goods that are non-critical yet very much in demand. Examples include recorded music, movies, foods that add pleasant variety to the local diet, books, more stylish items of clothing, sports equipment, cosmetics, perfume – for that matter, nearly all goods in which there is a flourishing trade in peaceful times. For interdependence to play its role in increasing the strength of the international system most effectively, it cannot be restricted only to goods that are of marginal interest to the economies involved. A flourishing trade in trivial goods will do little to bind nations together because the mutual benefits of such trade are minimal. Essentially then, a widening and deepening of international trade should be encouraged, with the exception of a relatively restricted set of especially critical goods.

The economic theory of comparative advantage holds that the world total of goods and services available is greatest if each nation specializes in those products it is best at producing, then trades for those goods and services it does not produce. It makes sense that a system in which everyone concentrates on what they do best and depends on others for the rest of what they need is potentially more efficient than a system in which everyone tries to be self-sufficient. Specialization and trade is in fact key to the internal *modus operandi* of modern economies.

The strategies recommended here preserve the spirit of this theory. But there are two significant departures. First, critical goods are exempted for political and psychological reasons. Second, the theory of comparative advantage is really more appropriate to a static than a dynamic view of the world. Developing the full economic potential of a nation inevitably will change the structure of its economy in ways that are likely to alter the pattern of comparative advantage. It is crucial to make room for such change. To require that each nation remain frozen in its present pattern of relative advantage is to hinder, if not to checkmate, its prospects for economic development. Establishing a peacekeeping international economy requires encouraging, not frustrating, the process of economic development. Moreover, it is best if the economic interdependence of nations involves multiple sources of supply and multiple markets. A web of multilateral dependence will tie more nations more closely together than will a set of disconnected bilateral relationships. The continued economic health of each nation will come to depend not only on the well-being of its own trading partners but also on the welfare of those nations that trade with its trading partners. Broader and stronger incentives for maintaining the peace in the face of conflict will therefore be created.

Of course, though generally it is wise to emphasize positive economic incentives, situations may arise in which negative economic sanctions (*e.g.*, boycotts) may be useful (*e.g.*, in instances of gross violations of human rights); and the question therefore arises: Would a web of multilateral dependence make such sanctions less effective? It is true that a web of multilateral dependence would render each nation less vulnerable to sanctions imposed by any other *single* nation. But widely supported and concerted *multilateral* sanctions can still be effective. Indeed, when sanctions are applied multilaterally, they are likely to be more powerful economically, politically, and psychologically. Moreover, the need to convince other nations to go along will help to prevent sanctions from being applied too casually and too often.

Strategy II: Balance Relationships

For reasons discussed earlier, maximizing the peacekeeping capacity of the international economy requires creating a situation in which the flow of benefits in every international trade relationship is as balanced as possible. Perfect bilateral balance is not required. It only is necessary that enough benefit flows in both directions to create the perception that neither side is being exploited by the other, and that continuation of the relationship on more or less the same terms is clearly in the interest of both parties.

Economists typically assume that all trade transactions are voluntary and that all parties are rational maximizers of their own well-being. Therefore, the very existence of trade between nations demonstrates mutual gain, because rational traders would not voluntarily exchange anything unless what they got was worth more to them than what they gave up. But these assumptions do not imply that gains will necessarily be balanced.

For example, if a person buys a piece of land for $70,000, an economist would assume that both the buyer and seller gain from the exchange—or they would not have agreed to it. But it is perfectly possible that the land is worth $60,000 to the seller (who therefore gains $10,000 from the trade) and $2,000,000 to the buyer (who therefore gains $1,930,000). Furthermore, even though both have gained, *if* the seller later learns that the buyer would have been willing to pay almost $2,000,000 for that land, it is likely that the seller will feel angry and exploited, poisoning future relations between them. If the transaction had taken place at a price closer to $1,030,000, on the other hand, both buyer and seller would have gained somewhere around $970,000 from their trade. Gain would be balanced and future revelations would be less likely to cause either party to feel angry or exploited (which of course is not to say that the buyer would not have preferred to pay the lower price).

Differences in market power, information, and bargaining ability can easily result in an unbalanced flow of benefits. It would be useful if there were some way of objectively establishing the degree of balance in trade relations to guide this strategy. There has been some attempt to develop a sensible, objective standard, but there is still much that remains to be done.[15]

Balanced gain, however, is only one dimension of balanced relationships. It also is important that input into the decision-making process as well as decisional power itself be balanced. If the decision process relevant to the relationship is lopsided, one side may feel that it is excessively dependent on the good graces of the other—a situation not conducive to feeling secure. On the other hand, a relationship in which both parties are equal partners in decision-making gives each of them the feeling of ownership in the relationship. It is theirs, something that they have created, not simply a gift that one of them unilaterally bestowed on the other and can just as easily withdraw. Both parties therefore tend to feel a responsibility for the "care and feeding" of the relationship, an obligation to help it succeed and to continue. Such feelings cannot help but strengthen the system.

Whether at the level of international treaty negotiation or interpersonal conflict mediation, imposed solutions – no matter how clever – are rarely as effective in dispute resolution as solutions to which both parties have contributed. If the parties themselves have participated in the creation of the agreed solution, they will tend to have a better understanding of the underlying issues and more confidence in the fairness of the settlement. Even if neither party is thrilled with the final result, there will be a greater tendency for each to abide by it. Also, and just as important, they will be more confident that the process of mutual interaction that brought the conflict to a successful resolution is capable of resolving future conflicts as well.

Strategy III: Create Flexible, Development-Oriented International Financial Institutions

Financing mechanisms can reduce strain and increase strength or add to strain and reduce strength. Properly structured financial mechanisms transfer present claims on economic resources (in the form of money) from those who have accumulated them to those who can put them to productive use. The smooth functioning of financial institutions thus prevents situations in which those with excess funds to invest cannot find fruitful uses for those funds, while those who have profitable opportunities for investment are not able to gain access to sufficient funds to take advantage of them. However, financial mechanisms can be used also as a tool for domination and control. Readily available credit on terms that make it easy to borrow is very seductive, particularly to those who can least afford it. An individual, company, or nation that gets itself too deeply in debt will soon discover that it has compromised, if not completely lost, its independence. The colossal expansion of international debt in recent decades has left the sovereignty of more than a few nations in question. This is particularly (although not exclusively) true of the world's less developed nations. By 1985, the external debt of the Third World was greater than $750 billion, more than 330 percent higher than it was in 1975.[16] As these nations have fallen more deeply into debt, they have been forced to make very difficult and painful choices.

The rising cost of servicing the debt began to absorb much of hard currency export earnings of these Third World nations, limiting their capacity to import both investment goods needed for development and consumer goods needed to maintain present living standards. To the extent that this interfered with the development process, the very growth of debt actually reduced their ability to generate the income needed to continue servicing the debt. To default on their loans would have created considerable international ill will as well as making it extraordinarily difficult to borrow in the future. Some Third World debtors had to borrow still more money just to make the required payments on their present loans. But given their increasingly fragile economic condition, international bankers were unwilling to lend additional funds to them unless they agreed to such draconian measures as wage freezes, devaluation of domestic currencies, and removal of government subsidies on food and other vital goods. Although such measures may be viewed as critical correctives by orthodox conservative banking principles, they deliver a severe economic shock to the local population, sending the price of many necessities soaring at the same time they prevent wages from rising. In more than a few cases, the result has been considerable domestic unrest, even revolution.[17]

For the government of any nation to be put into the position of having major domestic policy measures dictated by foreigners is a bitter pill to swallow. In the case of former colonies, this compromise of national sovereignty has led, not surprisingly, to cries of "recolonization." Whether this situation is the result of conscious attempts at domination or the best of intentions gone wrong, whether it is the fault of the more developed countries, the less developed countries, or the international bankers, it has created a great deal of real hostility, conflict, and insecurity. And that is something that a peacekeeping economy must avoid. Furthermore, deepening international debt, combined with less than stellar economic performance on the part of many debtors, also has made real the prospect of widespread default. Such a possibility is a serious threat to the stability of the international banking system, and thus to the economies of many nations, both rich and poor.

How, then, is it possible to construct sound international financial mechanisms that provide the economically critical lubricating function of finance while minimizing its conflict-generating tendencies? There are perhaps four basic principles.

1. *Maximize the extent to which debt financing is targeted to projects capable of generating the additional wealth needed to pay off the debt.* Many nations have borrowed enormous sums of money internationally to finance a variety of expenditures that have no prospect of generating additional economic wealth. This of course

means that repayment of the debt causes a net reduction in economic well-being, as existing wealth flows out of the country.

Debt-financed military expansion is one of the most common examples. In the early 1960s, the less developed countries as a group accounted for about 10 percent of world military expenditures. By the early 1980s, they accounted for more than twice that large a share. If the Third World's *share* of world military spending had been the same over the first six years of the 1980s (*i.e.*, 1980-1985) as it was in the early 1960s, the savings from this one source alone would have amounted to more than half of their outstanding 1985 debt.[18]

2. *Tie the fortunes of the providers of the capital to the success of those particular projects.* As long as the borrower (or others who stand ready to pay off the debt should the borrower default) has accessible sources of wealth sufficient to pay a loan, the lender has little incentive to worry about the viability of the particular project being financed. It is better to strengthen the incentives of the providers of capital to see that the project at hand succeeds. Joint ventures are one effective means of accomplishing this.

3. *Limit the burden of debt financing on borrowers.* This will both create a greater balance of benefit and avoid conflict-generating resentment on the part of the borrower, who might otherwise come to see repayment of the loan as yet another source of exploitation. Limiting the debt burden also allows borrowers a better chance of getting out from under debt associated with unsuccessful projects.

4. *Provide outright grants or heavily subsidized long-term loans for long-term public projects that are critical for development, but are not profitable in the narrow business sense.* There are many projects that would fail ordinary accounting criteria of profitability that are nonetheless worthwhile – even vital – to the success of development because they generate crucial social benefits that are difficult or impossible to measure. Infrastructural projects such as road building, improvement of sewage treatment and water supply facilities, and construction of schools often have this character. Ordinary debt financing of projects of this sort is inappropriate by the first principle and could well be discouragingly burdensome.

Strategy IV: Minimize Ecological Stress

Much of the world's ecological stress is the result of the economic activities of production and consumption. A high level of transnational ecological stress increases strain and is, therefore, inconsistent with a peacekeeping international company. Continuing international conflicts are created by, for example, chronic problems like acid rain. Acute environmental problems such as those caused by the breakup of an oil tanker or the meltdown of a nuclear power plant periodically raise the level of international confrontation still higher. Transboundary pollution by itself may not lead to war, but there is little doubt that it already has generated a degree of conflict and hostility and that it has the potential for generating a good deal more. According to Michael Renner:

> Massive damage from acid deposition in Canada (more than 50% of which comes from U.S. sources) has caused considerable diplomatic friction between Canada and the United States. . . . The Danish parliament decided to ask Sweden to close a [nuclear power] plant 30 kilometers from Copenhagen. The French government rejected a similar plea by local West German authorities to cancel construction of four reactors. . . . Tensions on similar issues run high between Ireland and the United Kingdom, between Austria . . . and both West Germany and Czechoslovakia, between Hong Kong and China, and between Argentina and Chile.[19]

Every additional source of tension contributes to the strain in the international system and thus to the likelihood that other sources of conflict will lead to the eruption of violence. The greater the load on the camel's back, the more likely the next straw will break it.

Perhaps even greater political conflicts have been generated by another form of ecological stress: the depletion of vital non-renewable natural resources. In fact, as discussed earlier, it has been argued that the desire to gain (and often, to monopolize) access to raw materials has been a significant cause of war historically. There is little doubt that it was one of the driving forces behind the colonization of much of the world by the more economically and militarily advanced nations. And though the imperial era has passed, these pressures continue to bring nations into conflicts of the most dangerous kind—conflicts in which each party believes that what is a stake is the continued economic well-being and perhaps political sovereignty or survival of its own people. It is hard to imagine that the Middle East, for example, would continue to inspire as much of the attention of the major military powers as it does today if it were not a source of critical supplies of crude oil. Such conflicts create considerable strain and thus a compelling threat to international peace. They must be minimized.

It has been argued that the expansion of economic activity itself is inconsistent with the maintenance of environmental quality, that modern activities of production and consumption inevitably generate a degree of ecological stress. There is an element of truth in this. Yet the levels of economic well-being to which populations of the more developed countries have become accustomed can be maintained, improved, and extended to the less developed nations without generating even the levels of environmental degradation we currently are experiencing. Doing so, however, requires (1) a great deal more attention to the efficient use of natural resources; (2) the development and use of pollution-abating technologies and procedures; and (3) a shift toward qualitative rather than quantitative economic growth.[20]

CHARACTERISTIC STRUCTURES AND INSTITUTIONS OF AN INTERNATIONAL PEACEKEEPING ECONOMY

There are a great many different specific structures and institutions that are consistent with the four basic strategies of a peacekeeping economy considered above. What follows is intended as illustrative rather than definitive, a first attempt to outline one feasible structure.

Generally Free Trade

A peacekeeping international economy would be characterized by relatively few barriers to trade. As long as trade relations are mutually beneficial, a large volume of trade will help bind nations together. In a world of mutually beneficial trade relations, all nations that are potential trading partners will tend to look upon tariff barriers and other such obstacles as a nuisance. The primary exception to the pattern of free trade would lie in the area of critical goods.

Tariffs and other trade barriers may be useful tools for encouraging greater national independence in critical goods, especially in the short run. If it is possible to arrive at general international agreement concerning what goods may be considered critical, the erection of barriers to trade in such goods will be understood as limited and will not result in a widening pattern of retaliatory tariff actions by other nations. However, although very useful, general consensus on what constitutes a critical good is not crucial. Failure to come to such agreement would simply leave each nation to concoct its own operating definition and to deal individually with any antagonism its definition might create among its present or potential trading partners. This would be less than ideal but still workable.

Other approaches to achieving independent critical goods supply capability may be less contentious than encouraging domestic production behind a protective wall of tariffs and other trade barriers. Direct government subsidy of domestic critical goods production may be cheaper, and less likely to encourage retaliation. Contingency plans for emergency production of critical goods, as in the Swiss example cited earlier, are even less likely to cause problems. They do not interfere at all with trade under normal conditions. Strategic stockpiling of storable goods is even less likely to generate international hostilities because it actually encourages the purchase of such goods, especially during the initial buildup of stores.

The widespread use of critical goods trade barriers might seem to make life especially difficult for those nations whose main exports are critical goods. But it is highly unlikely that all nations will be able to achieve complete independence in critical goods, at least in the short term. There is very little chance that a country like Japan, for example, will achieve independence in critical raw materials or energy in the foreseeable future, given that nation's relatively poor endowment of depletable natural resources. The contingency plan and stockpiling options are still open to such countries, but they would not interfere with the sales of critical goods exporters.

Furthermore, as it is necessary for any nation to be capable only of meeting its *minimum* critical needs independently during prolonged trade disruptions, there is plenty of room for trading internationally in normal times. Such trade is encouraged by the desire to increase the quantity, quality, and variety of critical goods available. Food is a particularly clear example. A nation may work to develop its capability to meet the minimum nutritional requirements of its population from local grain production or from stockpiles during emergencies, but under normal conditions access to imported fruits, meats, vegetables, and processed food products will make its population's diet a great deal more pleasant and interesting.

In general, the peacekeeping economy would be characterized by relatively free international trade in which multiple sources of supply would be the norm. Multiple sources are especially important for critical goods, but should be the basic structure of all trade (for reasons discussed earlier).

One way to achieve such a trade pattern would be through a combination of regionally integrated international free trade zones, like the European Economic Community (EEC) and broader interregional trade arrangements. The establishment of regional common markets has two potential advantages. First, it is smaller and easier to manage than a quantum leap to a truly integrated, free-flowing world marketplace; yet at the same time it is consistent with movement in that direction. It is thus a logical interim step. A pattern of gradually widening regional common markets is a reasonable, evolutionary path to global free trade. Second, to the extent that they are successful, regional economic communities create a base of cooperative experience and mutual gain that can lead easily to a higher degree of political as well as economic cooperation. The success of the EEC has led to the establishment of a fledgling European Parliament and a plan to eliminate virtually all remaining intra-regional trade restrictions by 1992. It also recently has given new credence to proposals for a single European Central Bank and a unified currency for Western Europe— proposals with profound political as well as economic implications.

Greater Trade and Economic Cooperation Among the Less Developed Countries

It is easiest to establish balanced relationships between parties of more or less equal power. Accordingly, greater cooperation among the less developed countries (LDCs) can help to break down the colonial mentality and patterns of dependency that are the legacy of the subordinate role established in the age of empire. Of course, it is misleading to suggest that all LDCs are at a comparable level of economic and political development or even that there is necessarily a great deal of homogeneity among them. However, LDCs are on the average closer to each other than to the more developed countries (MDCs). And for the most part the paradigm of dominance is not as historically entrenched in their interactions as it is in LDC-MDC relations. For both those reasons, it should be easier to establish mutually beneficial, balanced LDC relationships compatible with a peacekeeping economy.

A number of regional LDC free-trade zones have been established, with varying degrees of success. The Central American Common Market (CACM) was formed by five nations in 1960 and disbanded as a result of political differences in the 1970s. Fifteen countries with a combined population of about 124 million established the Economic Community of West African States (ECOWAS) in 1975. It continues to function today, but it cannot be said to have been an unqualified success.[21]

Of course, greater trade among the LDCs does not depend upon the formation of regional free-trade zones. There are good economic reasons why a general broadening of trade among the LDCs seems worthwhile. Third World producers should better understand the context and constraints of societies at lower levels of development than do their MDC counterparts. They should therefore be more attuned to manufacturing products whose design features, economic characteristics, and operating and maintenance requirements are more appropriate to LDC customers.

There are supply-side advantages as well. LDC manufacturers who strive to service MDC markets must meet more rigorous MDC standards for product quality. But quality is not cost free. Insisting on products of high enough quality to satisfy wealthier MDC customers typically will require a production process so expensive that it will price those products completely out of reach of all but the elites in the LDCs. Cheaper modes of production are likely not only to be more appropriate for LDC producers but also to produce goods that can be sold at prices that will make them accessible to a much larger fraction of LDC populations.

There are potential advantages to expanded LDC cooperation in marketing primary products such as food and raw materials to the MDCs. The success of price actions begun by the Organization of Petroleum Exporting Countries (OPEC) in 1973 is a spectacular example of the ability of producer coordination among LDCs to achieve prices for their primary products that more closely reflect their value to MDC customers. It would have been wiser for the OPEC nations to have raised prices more slowly to avoid the sudden shock and consternation their actions created. And it is a great pity that the gains to the OPEC nations were not used to mitigate the oil price shock on other LDCs and for more developmentally oriented purposes at home. Nevertheless, OPEC's success in shifting the international flow of benefit was impressive.

It may seem odd to argue that more monopolistic behavior on the part of the LDCs might be a good idea. After all, one of the central points of this analysis is that a peacekeeping economy must move away from the kind of exploitative behavior for which monopolies are famous. But the LDC producers in these markets often find themselves facing off against powerful and sophisticated MDC governments or multinational firms with which they have very little chance of striking an equitable bargain. Such a market, dominated by powerful monopolistic forces on one side, can move closer to a more equitable and efficient competitive outcome either by breaking up those forces or by increasing the power of other players in the market to countervail that power. John Kenneth Galbraith has suggested that there might be circumstances in which the latter approach could have advantages over the more traditional antitrust actions.[22]

This is clearly not an ideal situation. There is, for example, the very considerable danger that powerful LDC cartels will cooperate with powerful MDC-based multinational corporations rather than facing off against them, to the detriment of both MDC and LDC consumers. Worse yet, LDC cartels that abruptly force prices as high as did OPEC in the mid-1970s may create so much hostility as to increase strain in the international system. Therefore, it is worth giving careful attention to the search for other practical mechanisms capable of better balancing the international flow of benefits without resorting to the cartelization of LDC primary product exporters.

Special Purpose International Funds

There are three special purpose international funds that would be useful in achieving or at least facilitating the objectives of a peacekeeping international economy: a Global Environment Fund (GEF), a Global Infrastructure Fund (GIF), and a Global Bank.

Global Environment Fund

There are cases in which the consumption or production activities of one nation generate pollution that causes ecological damage abroad. Acid rain is a case in point. Suppose Country A is creating pollution that results in acid rain falling on Country B. Substantial improvement in environmental quality in Country B can be achieved if Country A either halts the pollution-generating activity or institutes antipollution measures. If Country A ceases that activity, it loses the benefits associated with it; if it cleans up the activity, it bears the costs of the cleanup; and, in either case, the main environmental advantages accrue to Country B. Operating out of narrow, short-term economic self-interest, Country A has no incentive to cease polluting the environment of Country B. Now, suppose Country A is an LDC whose resources are already stretched to the limit in its attempt to meet its own needs and to get sustainable development going, while Country B is an MDC in a relatively strong economic and technical situation. Would it not make sense for Country B to offer assistance to Country A to help in the abatement of this pollution?

A Global Environment Fund could provide grants to LDCs to assist them in undertaking projects with strongly positive environmental impacts. These might involve improved sewage treatment facilities, air pollution control measures, protection of endangered species, or preservation of rare environmental assets such as rapidly disappearing hardwood forests. Where it is ideologically acceptable, the GEF itself might buy land in a given country and hold it as an environmental preserve. Or it might provide a "perpetually forgivable" loan to the government or to an indigenous group to buy the land and preserve it. Loan payments would be due to begin at a specified time each year, but at that time the due date would be postponed another year if the area still remained untouched. As long as the area purchased remained an environmental preserve, no loan payments ever need come due. Thus, the Global Environment Fund would help to institutionalize Strategy IV of a peacekeeping international economy, the minimization of global ecological stress.

Global Infrastructure Fund

The GIF would be similar to the GEF in its structure and function but would finance projects to improve Third World systems of transportation, communication, and water and energy supply. Such a fund has been proposed by Masaki Nakajima, an economist who headed Japan's Mitsubishi Bank. The Nakajima "global Marshall Plan" would be financed by the MDCs, making available $25 billion per year as seed money for mega-projects "that would have worldwide impact but are too costly to be undertaken by a single nation, such as massive water projects to 'green' the world's deserts."[23] The problem is, however, that the economic impact of such "macro-engineering" is likely to be more positive for the MDC construction firms that build them than for the Third World. Furthermore, the environmental impact of such grand projects is questionable.

What is needed instead is a grant-making fund that finances a variety of infrastructural projects, many of them small or moderate in scale. For example, the digging of many individual wells may be a more effective and environmentally sound approach to irrigation in some places than building huge dams and pipelines. Building a network of relatively simple roads may be more a contribution to development than building modern superhighways that span continents. Large-scale projects may sometimes be desirable, but the highest probability is that small- to medium-scale projects will make the most sense most of the time.

Perhaps the most critical area for the GIF is in improving the quality and availability of potable water. It has been estimated that more than three-quarters of the disease in the LDCs is traceable to unclean water. Insufficient water also threatens food supplies. Drought, along with a great deal of political disruption, has brought, for example, mass starvation in recent years to Ethiopia, Somalia, and the Sudan. Projects to address this critical need are a natural area of focus for the GIF and cooperation between the GIF and the GEF.

Thus, the GIF would be responsive to Strategy III, *i.e.*, providing a financial mechanism less likely to generate hostility and cries of economic recolonization than present systems. Also, it would encourage real economic development, not just economic growth (*see* this chapter: Sustained Multidimensional Efforts in Development).

Global Bank

A Global Bank committed to providing financing and technical assistance for development projects is an extremely good idea also. The present World Bank may offer a useful starting point. Unlike the GEF and GIF, the Global Bank would not provide outright grants primarily but, rather, loans and limited-return equity financing.

Care must be taken to avoid creating an insupportable level of debt that interferes with, rather than encourages, development. Some low-interest, long-term development loans should be provided, but the Bank should emphasize equity financing for particular development projects with profit potential. In effect, the Bank would buy the equivalent of non-voting preferred stock in the project. If the project became profitable, the Bank would receive limited dividends, as is typical of preferred stock. At any time, the LDC organization involved could buy back this preferred stock at its original price. With such financing, unsuccessful projects would not create a continuing burden for the LDCs, yet successful projects would not see a disproportionate share of their return continuing to leave the country. The LDCs' downside risk consequently would be limited without unduly restricting the profit incentives of LDC investors.

Thus, the Global Bank, like the GIF, would be an institution responsive to Strategy III. It would help in the creation of flexible, development-oriented international financial mechanisms and procedures assisting in real economic development (*see* this chapter: Sustained Multidimensional Efforts in Development).

In sum, the GEF, the GIF to lesser extent, and the Global Bank, each unsustainable in the absence of continuing external support, would not be profit-making institutions but, rather, mechanisms through which the economically stronger MDCs could provide aid to the LDCs on a multilateral basis. If MDC participation were broad enough, it would be possible for these institutions to reduce or avoid the more blatant political manipulation and abuse that so often has accompanied bilateral aid in the past. The primary MDC incentive to offer this aid would be that it would constitute a more effective and cheaper way of buying security than continuing to pour huge sums into oversized, insatiable militaries. Other incentives would include improved environmental quality and greater economic development with the potential for expanding markets and improving the quality of life of MDC as well as LDC populations.

Sustained Multidimensional Efforts in Development

Economic growth is simply a matter of expansion in the size of the economy. The fact that an economy is growing does not by itself assure that the material condition of the population is improving. Economic development is a complex matter of qualitative as well as quantitative change in skills, productive capacity, level of output, health, nutrition, education, organization, and general material well-being. Real development raises capability, provides opportunity, and improves living standards, not just for a privileged few but for the broad mass of the population. Until there is a great deal more real economic development and not merely economic growth in the Third World, it will be difficult for the peacekeeping economy to reach its full potential as a source of global security. It is much easier and more natural for the kind of mutually beneficial relationships that characterize the European Common Market to arise and persist among nations at similar levels of development, high enough to render them valuable to each other as customers and suppliers. If they were all comparably poor, they would have much less to offer each other, which is one of the reasons why some present and past LDC attempts at common markets have not been as successful as they might otherwise have been.

Generating sustained economic development is one of the most important elements in creating greater security through a peacekeeping economy. A rising standard of living tends to reduce the desperation people feel when they are hard pressed to provide themselves with the material necessities of a decent life. True, those living in the most extreme conditions of poverty and deprivation are unlikely to have the strength or the means to engage in sustained social violence. All of their energies are expended in a constant fight to survive. But a great many of the world's poor are not at the barest edge of survival, and they have strong incentives and at least minimal means with which to struggle for change. They are much more likely to turn to extreme solutions and generally are more willing to support, even to participate in, disruptive violence than people in better economic condition. They simply have fewer alternatives and less to lose. Witness the Palestinian-Israeli dispute.

Although not a panacea, real economic development will remove some important barriers that otherwise block the achievement of a peacekeeping international economy. A world of developed nations is not inherently a world without war, but it would be a world in which the poor would be less desperate and the rich less frightened that "the great unwashed masses" would abruptly pull them down from their mountaintop of privilege and power.

This fear of the rich motivates much of the forcible, often violent, suppression of the poor. But suppression increases the frustration, anguish, and desperation of the poor, which in turn creates and legitimizes greater fear. Locked in this cycle, there is little chance to prevent permanently the eruption of violent confrontation and avoid a widening war that a world armed with modern weapons of mass destruction simply cannot afford. It is true that the conflict between the rich and poor is at least as much a conflict between the elite and the impoverished people of the LDCs as between those who live in the richer countries and those who live in the poorer countries. But we certainly have seen enough examples in this century alone to understand that struggles that begin within Third World nations can jump beyond their borders and/or escalate into wars in which major military powers become directly or indirectly involved (*e.g.*, the civil wars in Afghanistan, Angola, Ethiopia, Nicaragua, and Vietnam). Real economic development can contribute to world peace and security by helping to break this cycle.

There is no technical reason and certainly no moral excuse for the persistence of widespread absolute poverty in a world as productive as that in which we presently live. The elimination of absolute poverty is in our grasp and in our interest. It is an achievable, security-enhancing goal. Beyond this, sustained Third World development should further reduce the likelihood of war by softening the economic pressures on the wider LDC populations. Also, nations have more to lose by disruption of trade as their economic condition improves. And it is easier to establish mutually beneficial relationships between nations as the gap between them narrows.

In a sense, the ultimate goal of global development is to eliminate the Third World as an identifiable entity. Although international differences in living standards may never be completely eliminated, there is no reason why there must continue indefinitely to be a permanent economic underclass of nations. Given this ultimate goal, some of the institutions discussed above can be seen as transitional rather than permanent. If it does its job well, the Global Bank, for example, should eventually put itself out of business, or at least evolve into a very different institution. The GEF and the GIF should likewise either disappear or mutate into agencies for liaison and coordination of global efforts in these areas, rather than as grantmaking institutions *per se*. The value of emphasizing intra-LDC trade should also fade and finally vanish as real development succeeds on a global scale. The success of these extremely useful interim approaches will be measured by how quickly and completely they render themselves obsolete.

International Regulation of Multinational Enterprises

Multinational corporations (MNCs) represent an evolution of business beyond the evolution of political systems. Their ability to coordinate financing, investment, production, and marketing activities worldwide is unparalleled by the reach of any single political authority.[24] To some, they are the most dangerous and powerful exploitative force in the world today. To others, they are the greatest force for encouraging development and international economic relationships that build trust and understanding worldwide.

Multinational business already has created a vastly more integrated and interdependent international economy and has the potential to take this process much further. That is to the good. The problem is that much of this integration has been unduly one-sided, resulting from decisions made in pursuit of corporate goals emphasizing the growth of the economic power and size of the corporation as an end in itself. However, there is no necessary reason why MNC decisions could not be geared to the pursuit of the broader and longer term goals that characterize the peacekeeping international economy.

The extent to which MNCs have, on balance, encouraged or discouraged real economic development is still a matter of controversy. Have they contributed additional capital, trained more indigenous personnel, and transferred useful technology to the LDCs? Or have they drained limited LDC capital, pre-empted already skilled workers, and transferred only technology that is so sophisticated and inappropriate to LDC conditions that it has little positive spinoff to the rest of the economy?[25]   Either way, it is important to understand that the effects on development, whether positive or negative, have been incidental to the primary goals of the MNCs.   MNCs are not in the business of encouraging or discouraging development – they are in the business of making money.   However, it is clear that in some cases at least, they *have* interfered with development directly and/or actively supported repressive governments more interested in law and order than in justice or development. To the extent that MNCs have done this, they have worked in a direction contrary to the purposes of a peacekeeping international economy.

When the narrowly focused interests of private enterprises lead them to take actions that are contrary to the broader public interest, the best way for governments to realign the divergent interests is to restructure the rules of the game under which those enterprises must operate.   Ordinarily, this is much cheaper and more effective than trying to intervene in a direct and heavy-handed way in the organization's internal decision-making.   For example, suppose a firm were polluting the air with noxious sulfur compounds. The direct interventionist approach might call for a law requiring the firm to burn only low-sulfur fuel oil. But it might actually be much less expensive to burn high-sulfur fuels and use filters to remove the sulfur in the smokestack gases before they leave the factory or possibly to burn a low-sulfur fuel other than oil. Those options and their relative costs are more likely to be known to the firm whose business is at stake than to the government. The restructuring approach, however, might be to institute a fine based on the amount of sulfur that is allowed to escape the stack and to let the firm worry about how to avoid it. This approach changes the rules of the game so that the firm must internally bear the cost of polluting. But it leaves maximum flexibility and freedom of action in the hands of those most able to find an efficient solution to the problem.

To the extent that MNCs are operating at cross-purposes to the principles of a peacekeeping international economy, rules of the game must be established that will cause them to shift to a more desirable behavior without direct intervention. A single world government capable of regulating MNC behavior is not required. It is perfectly possible to achieve a similar result by international cooperation established through mechanisms such as a generally agreed convention on the conduct of multinational business. There are, for instance, a series of such agreements governing the conduct of international air travel, postal service, and behavior on the high seas. To be sure, these agreements are sometimes violated, but on the whole compliance is quite good. Enforcement could be achieved by domestic legal arrangements in each signatory country (the equivalent of multilateral extradition treaties), combined with economic sanctions. Such sanctions are likely to be even more effective against economic enterprises than against renegade governments.

Regulation of International Trade in Hazardous Materials

Because of differences in health and safety regulations among countries, it is not uncommon for MDC manufacturers to sell unsafe or at least questionable products to the LDCs that cannot lawfully be sold domestically—for example, pesticides and pharmaceuticals. Such trade, which sends a clear message that people in the LDCs are inferior in the eyes of the MDCs and that they are undeserving of respect for their health and well-being, is not only profoundly immoral but generates the kinds of international antagonism and conflict that a peacekeeping economy must seek to avoid. It is foolish and counterproductive in other ways as well, often exposing MDC consumers of LDC imports to the very dangers their countries have sought to avoid.[26] Therefore, it makes sense to prohibit international trade in such materials; and probably the simplest way is by international agreement, requiring each nation to write into its own legal code a prohibition on the export of any product or material that cannot legally be sold within its own boundaries. Such a law could include the activities of firms that operate outside the nation's political jurisdiction as subsidiaries of firms based within its jurisdiction.

There is a flourishing international trade in another kind of dangerous material as well: finished weapons and the equipment, components, and materials critical to building them. Such trade also is antithetical to a peacekeeping international economy and likewise must be stopped. Trade in materials and equipment critical to the manufacture of nuclear, biological, and chemical weapons especially must be restricted, to the extent possible without interfering with non-hazardous civilian use. And except for very small arms, weapons themselves have no meaningful civilian purpose and therefore should be strictly controlled as well. The International Atomic Energy Agency (IAEA), a key element of the Non-Proliferation Treaty of 1968,[27] is set up as an international inspector, monitor, and regulator of civilian nuclear facilities and trade in nuclear materials. Underfunded and understaffed, it should be re-evaluated to see whether it makes more sense to strengthen and expand it or to replace it with a wholly redesigned organization. In any event, whether or not the weapons and weapons material trade encourages conflicts, it certainly increases the likelihood that the conflicts that do arise will explode into violence. It also dramatically increases the level of damage done when war does erupt, as well as the likelihood that suppliers of critical arms to the combatants will be drawn into the conflict. All of this is inconsistent with a peaceful world or a peacekeeping international economy.

**International Council on Economic Sanctions
and Peacekeeping**

Economic sanctions do not generally have a good reputation among specialists in international relations. However, I believe that sanctions can have a useful, though minor, role to play in a peacekeeping international economy.

To be effective, economic sanctions must have broad enough support so that they are not easily circumvented. Given the flexibility of trading systems, this is not easy to achieve. For example, when the Carter Administration instituted an embargo on grain sales to the USSR in response to the Soviet invasion of Afghanistan, Moscow simply bought the grain it needed from other sources. If the United States had managed to enlist the cooperation of all the other major grain exporting nations, the Soviet Union would have had a more difficult time. Of course, as long as some other nation that could buy grain from the exporters would have been willing to resell to the Soviets, the embargo would have failed anyway. But if the embargo had been supported by all the major grain exporters, probably the Soviets would have had to pay significantly more for the grain and thereby suffered some economic penalty.

Furthermore, economic sanctions tend to work more slowly than other punitive approaches. Thus the sanctions so far instituted against South Africa have not yet resulted in the abolition of apartheid. However, if these sanctions are as ineffective as some claim, why does the Government of South Africa go to so much trouble to convince people that they are ineffective? If they really are ineffective, would it not be more sensible for South Africa to pretend that they did hurt so that their opponents would be less likely to switch to more effective means of punishment? Economic sanctions create pressure, and that pressure often works slowly. Unless the sanctions are extraordinarily broad and widely supported, they are unlikely to serve as a quick fix. But as an adjunct to positive incentives and other forms of pressure, they still can be useful.[28]

There are three basic categories of economic sanctions: trade embargoes (including the erection of very high, punitive tariffs), financial boycotts (including divestment, disinvestment, and denial of foreign aid), and the freezing of assets abroad. In a global economy embodying the principles discussed in this chapter, it would be exceedingly difficult for any nation to impose an effective trade embargo unilaterally. Critical goods independence and an emphasis on development of multilateral trade provide too many alternatives. Unilateral financial boycotts are also almost certain to have limited effect and for similar reasons – with perhaps one odd but notable exception. If one nation has borrowed a great deal from another, the nation in debt may be able to apply considerable pressure to the creditor nation by threatening to cease repayment of the loans. If its debt is large enough, that sort of boycott could pose a real threat to the financial stability of the creditor. Of course, a debtor nation taking such an action might find it extremely difficult to borrow from *any* potential lender in the future. Unless it has overcome its need for external financing, its ability to carry out a unilateral boycott of this sort successfully is certainly limited. Likewise, a unilateral freezing of assets has a chance of succeeding only if the nation being subjected to the sanctions holds significant assets in the nation imposing the freeze. A nation would be especially vulnerable to such unilateral sanctions if, for example, its investors owned a considerable amount of real estate and other fixed capital assets (*e.g.*, oil refineries, chemical factories, automobile manufacturing plants) located in *one other nation*. On the whole, multilateral sanctions will be far more effective.

Effective multilateral sanctions require a more or less objective mechanism to trigger the sanctions, as well as an organization to decide which sanctions are most appropriate and to coordinate and monitor their imposition. A Council on Economic Sanctions and Peacekeeping ("the Council"), free of the veto that hampers the United Nations Security Council, could be established within the framework of the UN (with representatives from *all* the UN member States) or as an independent entity (with the same or similarly inclusive representation), complete with a permanent staff of conflict resolution specialists charged to monitor developing trouble spots, to bring the issues to the attention of the Council, and to make suggestions regarding the formal or informal procedures that might be most helpful in each instance. Any nation that has directly attacked the territory of another nation would be subject to, say, an immediate and total trade embargo imposed by the Council, and the embargo would continue until hostilities ceased and a UN military peacekeeping force could be deployed to monitor compliance with a ceasefire or withdrawal. The sanction would be automatic and mandatory whether or not the offending party is itself a member; and any violator of the sanction imposed by the Council, whether or not it is a member, would likewise become automatically subject to the same sanction until such time as it came into compliance. Further, any member of the Council could petition the body to take action. However, an overwhelming majority (say, three-fourths) probably would be desirable relative to the imposition of sanctions for any *domestic* or *internal* behavior (*e.g.*, the practice of apartheid by South Africa within South Africa).

How would it be possible for the Council to enforce its actions? In the same way, I submit, as any international regime is enforced. In the absence of overarching political authority (whether world government or hegemon), compliance with international agreements relies primarily on the value of the agreement to the parties involved. For example, though the 1968 Non-Proliferation Treaty previously mentioned makes no provision for punitive enforcement (indeed, it permits withdrawal at any time simply upon ninety days' notice), no signatory non-nuclear weapons State is known to have violated it. Why? Because of the benefits that are gained as a result of other State parties restricting their behavior in accordance with the treaty. Most international agreements rely to a significant degree on this process of mutual back-scratching. To be sure, the process does not always work. Treaty violations do take place, just as laws in our national societies are broken every day. But in a world more oriented to non-violent international peacekeeping than the one in which we live at present, a world in which progress toward a peacekeeping international economy is well under way, it is not unreasonable to expect that the Council would be seen as an institution with a sufficiently beneficial function to warrant active compliance and support. This might be particularly the case if the Council were authorized also to resort to positive inducements to resolve international conflicts (as did, for example, the United States at Camp David when it found it helpful to provide some financial assistance to Egypt and Israel as part of the conflict resolution process that led to the peace treaty between the two nations). The Council would therefore have the right to petition the Global Environmental Fund, the Global Infrastructure Fund, and the Global Bank to finance a particular project it proposes as a means of settling a potentially dangerous conflict.

CONCLUSION

The combination of (1) basically free trade, (2) tariff barriers (if necessary) to achieve independence in minimum critical goods requirements, (3) more LDC trade (including Third World common markets), (4) LDC primary product cartels, and (5) regulation of multinational corporations and trade in hazardous materials may seem an odd mix. It includes elements of free market, protectionist, interventionist, regionalist, globally oriented, and self-reliant development ideology. In a sense, however, this eclecticism is not surprising, as it arises from an attempt to strike two critical balances: a balance between independence and interdependence and a balance in the flow of benefits derived from international trade.

Clear, strong, and concrete emphasis on generating sustained progress in real economic development also is crucial to achieving this dual balance, most particularly over the long run. A Global Infrastructure Fund and a Global Bank could prove to be highly useful transitional institutions in the realization of this goal, while a Global Environment Fund, along with the regulation of international trade in hazardous materials, could help reduce stress on the environment, thereby minimizing ecological conflict. Agreements restricting international traffic in arms, though unlikely to reduce conflict itself, could sharply reduce the outbreak and extent of organized violence. In addition, the Council on Economic Sanctions and Peacekeeping could impose multilateral economic sanctions in such a thoroughgoing and coordinated way as to be far more punishing to violators of generally agreed norms of international behavior than present unilateral or ad hoc multilateral attempts. It also could provide expertise in conflict resolution and broker the use of certain positive inducements to settle conflicts amicably.

Is all of this realistic? Is there any reason to believe that it actually could come to pass? It is hard to imagine this world of fractious, violent, and heavily armed nations breaking the habit of nuclearism, war, and the threat of war.

Yet we are living in a time of profound change – in attitudes, technology, and in the political and economic conditions that surround the institutions of militarism and war. The end of the 1980s has seen remarkable change at sometimes breathtaking pace. In less than five years, relations between the US and the USSR improved enormously, as the Soviet Union underwent a revolution in openness and economic structure more dramatic than anything it had seen since the revolution that brought it into being. After a decade of essentially non-violent struggle, Poland broke through the rigidity and authoritarianism that had held that nation in its grip since World War II. In East Germany, that breakthrough took less than three weeks, and in Czechoslovakia only about ten days. Bulgaria and Hungary also experienced rapid and positive change. Except for the tragic case of Romania, the peoples of Eastern Europe, without raising a gun, without using any of the tools of war, began the process of liberating themselves from the straitjacket of socioeconomic and political oppression under which they had lived for so long. Though it is too early to tell how long-lasting these changes will be, the occurrence of transformations believed virtually impossible only a few weeks or months prior to their actually taking place makes it seem that anything is possible, given the appropriate political will.

Can we find the political will necessary to make nuclear deterrence and militarism and war curiosities of the past rather than threats to our future? I believe we can, I believe we must, and more than that, I believe we will.

NOTES

This paper was prepared with funding from the Exploratory Project on the Conditions of Peace (EXPRO). The author wishes to thank EXPRO for its generous support. The author also wishes to thank Kenneth Boulding, Mary Clark, Warren Davis, Dietrich Fischer, Bruce Russett, Burns Weston, and Ralph White for their many useful comments and criticisms.

1. For an exceedingly interesting presentation of this view, see anthropologist Richard Leakey's "Survival of the Species," the last program in his public television series *The Making of Mankind* (British Broadcasting Company Television in association with Time-Life Films, Inc., 1983).

2. The phrase is taken from Burns H. Weston, "General Introduction: The Machines of Armageddon," in Burns H. Weston (ed.), *Toward Nuclear Disarmament and Global Security: A Search for Alternatives* (Boulder, CO: Westview, 1984): 1.

3. For a discussion of the mercantilist and Marxist perspectives, see, *e.g.,* Stephen Gill and David Law, *The Global Political Economy: Perspectives, Problems and Policies* (Baltimore: Johns Hopkins University Press, 1988): chs. 3 and 5.   Lenin of course argued that imperialism played a key role in artificially postponing the demise of capitalism.   More recently, William Appleman Williams (of the so-called Wisconsin school of American diplomatic history) has argued that US foreign policy has sought to expand markets abroad, gain access to raw materials, and so on, to avoid the need to address serious issues of social and economic inequality at home.   *See* William Appleman Williams, *The Tragedy of American Diplomacy* (New York: Dell, 1972).

4. *See* Adam Smith, *The Wealth of Nations* (New York: The Modern Library, Random House, 1937): 314-15, 325-26.

5. *See* Lloyd J. Dumas, *The Overburdened Economy* (Berkeley: University of California Press, 1986).

6. *See* Gill and Law, *Global Political Economy*: ch. 4.

7. Richard Rosecrance, *The Rise of the Trading State* (New York: Basic Books, 1987): 24.

8. *See* Peter B. Kenen, *The International Economy* (Englewood Cliffs, NJ: Prentice-Hall, 1989): 218-20.

9. Karl Marx, *Selected Writings in Sociology and Social Philosophy* (Tom B. Bottomore and Maximillien Rubel translation) (London: Watts, 1961).

10. Adam Smith, *Wealth of Nations*: 581-82.

11. The shift from "scientific management" to the "human relations" school of thought (in the second quarter of the Twentieth Century) provides theoretical and empirical support for this proposition. Scientific management held that worker productivity could be maximized by: (1) breaking work down into tasks that workers were required to perform in a strictly specified way; and (2) manipulating the physical conditions of work (lighting, heating, etc.) and the system of monetary incentives associated with work in a wholly

authoritarian manner. The idea was to establish externally determined constraints that amounted to a system of coercion and control imposed on workers from outside. Beginning in 1927 with the Hawthorne studies, a body of empirical research developed that slowly overturned these ideas by demonstrating productivity was more effectively increased by establishing friendly relations among workers and an easier, less coercive supervisory style that generally made for a friendlier, more pleasant work environment. Studies by Elton Mayo, by Ronald Lippit, Ralph K. White, and Kurt Lewin, by Lewin and associates, and by Lester Coch and John French, Jr., help to found the human relations school. *See, e.g.*, Curt Tausky, *Work Organizations: Major Theoretical Perspectives* (Itasca, IL: Peacock Publishers, 1980): especially 41-53.

12. Charles Osgood, a psychologist specializing in communication, developed a strategy for reducing the level of tension in international relationships from the insights afforded by the study of interpersonal relationships. Osgood's extremely creative approach, known by its acronym GRIT (Graduated and Reciprocated Initiatives in Tension-reduction), emphasizes the importance of commitment and relies heavily on the use of unilateral initiatives to replace hostility and distrust with more cordial relations. *See* Charles E. Osgood, *An Alternative to War or Surrender* (Urbana, IL: University of Illinois Press, 1962). *See also* Charles E. Osgood, "Disarmament Demands GRIT," in Burns H. Weston (ed.), *Toward Nuclear Disarmament and Global Security: A Search for Alternatives* (Boulder, CO: Westview, 1984): 337-44.

13. *See* Kenneth E. Boulding, *Stable Peace* (Austin, TX: University of Texas Press, 1978): 33-36.

14. Dietrich Fischer, *Preventing War in the Nuclear Age* (Totowa, NJ: Rowman and Allanheld, 1984): 147.

15. In "Self-Organization, Transformity and Information," *Science* (November 25, 1988), Howard T. Odum argues that self-organizing systems, like the natural ecology and human society, involve competition among different subsystems. The subsystems that survive and prosper will be those in which the higher-order members (such as fish in a pond) provide feedback most efficiently to the lower-order members (such as algae), so that this feedback reinforces the production of what the higher order members need. Assuming a survival-of-the-fittest competition, and lots of time for trial and error, in the long run those processes that have survived will be those that use their inputs most efficiently. The most basic input is solar *emergy* (spelled with an "m" rather than an "n"). The solar emergy of any product — whether a pound of copper or a grapefruit — is the amount of solar energy *operating through existing real world systems of production* (and thus systems that are optimally efficient) needed to produce that product.

It is thus possible to value everything in terms of the solar emergy it embodies. The flow of goods involved in international trade can thus be evaluated to see whether the flow of emergy into one country from its trading partners is the same as the flow of emergy out of the country into its trading partners. This is an objective way of assessing whether trade relations are balanced. One problem with this approach, however, is that there are no grounds for assuming that human economic processes operate in a purely economic survival-of-the-fittest system or, even if they did, that they have operated long enough to have achieved long-run optimum. Therefore, the system of production in Country A may be far less efficient than that in Country B, making it seem as though Country A's products are worth more because they embody far more solar emergy. Odum's analysis is nevertheless a creative and interesting approach to the issue of balance and perhaps one that provides some basis for reasonable comparisons.

16. *See* Ruth Leger Sivard, *World Military and Social Expenditures, 1986* (Washington, DC: World Priorities, Inc., 1986): 20.

17. For an interesting analysis of the conditions placed on African debtor nations by the International Monetary Fund and their effects, see Magnus L. Kpakol, *The Political Economy of International Monetary Fund Conditionality Programs in Africa* (University of Texas at Dallas: Ph.D. Thesis, 1988).

18. *See* Ruth Leger Sivard, *World Military and Social Expenditures*, 1987-88 (Washington, DC: World Priorities, Inc., 1988): 42.

19. Michael Renner, "Enhancing Global Security," in Lester R. Brown (ed.), *State of the World, 1989* (New York: Norton, 1989): 142-44.

20. For a discussion of the role of energy conservation in this context, see Lloyd J. Dumas, *The Conservation Response: Strategies for the Design and Operation of Energy-Using Systems* (Lexington, MA: D. C. Heath, 1976).

21. Looking at the evidence of a few case studies, development economist Michael Todaro concludes: "Third World countries at relatively equal stages of industrial development, with similar market sizes . . . stand to benefit most from . . . economic integration. . . . In any event, . . . without cooperation and integration, the prospects for sustained economic progress are bleak. . . . It is quite possible that the pressures for some form of economic and political integration will gradually overcome the forces of separation and continued dependency." Michael P. Todaro, *Economic Development in the Third World* (New York: Longman, 1985): 426.

22. *See* John Kenneth Galbraith, *American Capitalism: The Concept of Countervailing Power* (Boston: Houghton-Mifflin, 1952).

23. Lee Dye, "Japanese Hint at Establishment of Ambitious 'Global Marshall Plan,'" *Los Angeles Times* (July 12, 1986): 25, cols. 1-3.

24. Though it has become somewhat dated, the classic book approaching multinational corporations from this perspective is Richard J. Barnet and Ronald E. Müller, *Global Reach: The Power of the Multinational Corporations* (New York: Simon and Schuster, 1974).

25. For an overview discussion of this question and some guidance to the relevant literature, see Gerald M. Meier, *Leading Issues in Economic Development* (Oxford: Oxford University Press, 1989): 263-67.

26. For one thing, pollution knows no national boundaries. A substance sold to LDCs that has been banned in an MDC because it has been found to be ecologically harmful may create environmental damage that will cross over

MDC-LDC borders. Furthermore, when MDCs import LDC products that have been produced with the aid of these banned substances, they may import the dangers as well, frustrating their own regulations.

27. Treaty on the Non-Proliferation of Nuclear Weapons, July 1, 1968, 21 U.S.T. 483, T.I.A.S. No. 6839, 729 U.N.T.S. 161.

28. The difficulties in enlisting wide enough support to give economic sanctions real bite are part of the reason why punitive measures such as military action that can be implemented unilaterally or with only limited allied support seem more appealing. Yet sanctions that are relatively ineffective without widespread support are generally more consistent with the spirit of a peacekeeping economy. Multilateralism in the international arena is a closer analog to democracy within the nation-state than unilateralism. Thus, Bruce Russett's arguments as to the value of the spread of democracy to international security may have some applicability here as well. *See* the chapter by Bruce Russett in this volume.

6

Psychology and Alternative Security: Needs, Perceptions, and Misperceptions

Ralph K. White

Our true nationality is mankind.
— H.G. Wells
The Outline of History

The general thesis of this chapter is that certain basic human needs – security, pride, and economic well-being – do not inherently promote war, but do promote war, in the present anarchic, Hobbesian state of the world, when combined with certain definable forms of misperception. Therefore, an immediate way to reduce the danger of nuclear war, before a world free of nuclear weapons can be achieved, is to cultivate the corresponding forms of realism (or accurate perception), together with progress out of the present state of anarchy.

If and when a post-nuclear world emerges, probably with effective forms of regional and world organization, and probably also with strict limits on conventional arms as well, the psychological situation will be different in ways that are well worth considering now. Of course, the basic human needs for security, pride, economic well-being, and community approval will continue to exist and to demand satisfaction as they do now. And, then as now, they will need to be combined with much realism, especially about one's own country and its potential enemies so as to promote continuing peace rather than war. But there also are likely to be important

176

psychological differences, most of them serving to reinforce the effectiveness of the regional and world organizations that then would be established. For instance, there probably would be:

1. *A much increased tendency on the part of individuals to identify more with the human race as a whole than with any of the nations that compose it.* Such identification, and the sense of loyalty that is likely to accompany it, would be a natural consequence of whatever forms of regional and world organizations are developed, as well as of many trends that are taking shape at the present time: worldwide communication, economic interdependence, and so forth.[1]

2. *A more realistic perception of one's own nation and of its potential antagonists, involving large departures from the stark Good Guys/Bad Guys caricature that presently is fostered by politicians, educators, and the mass media, all of whom realize that their rapport with their publics (or constituents) depends partly on making those publics feel good about themselves, with a general avoidance of any approval of their country's opponents.* The Good Guys/Bad Guys caricature is perhaps the chief psychological factor contributing to war.[2] But, though an unwholesome situation, it is unlikely to change in any fundamental way as long as the public to which a politician, an educator, or the mass media communicates is composed exclusively or almost exclusively of members of the same nationality. However, it could be changed significantly if the officials of future regional and world organizations are chosen by, and responsible to, transnational publics, and if they are in a position to promote objectivity in politics, education, and mass communication.

3. *A diminished importance of the highly competitive "macho" form of national pride.* Such pride is now fostered by the arms race and the unbridled national competition for power, prestige, allies, client States, buffer territories, and economic advantages. In a post-nuclear and largely disarmed world there still will be ample room for pride, including national pride, in other fields—economic productivity, human welfare, sports, science, and the arts—but not in military power.

4. *Greater economic well-being.* When the extreme economic wastefulness of war and preparations for war are largely ended, the economic benefits to the world as a whole will be potentially enormous. Freed from that burden, nations could cooperate more wholeheartedly in tackling the problems of overpopulation and environmental deterioration, problems that are likely to persist and grow in the next few decades regardless of what happens politically.

Such cooperation, in turn, could help reinforce the new harmony between nations.

5. *Less exclusive caring about the approval of one's fellow nationals, and more attention to the approval of the rest of the human race.* Philosophers often have marveled at the supreme egoism of nations. Reinhold Niebuhr wrote about "moral man and immoral society," and a main reason for it surely is the fact that people are continuously in contact with their fellow citizens and depend greatly on their approval while foreigners are, as a rule, shadowy and faraway creatures whose approval or disapproval is usually not known even if sought. The much more frequent contacts and greater knowledge of other countries that could be expected in a more integrated world could change this situation significantly. In fact, after some decades of successful operation of, let us say, an international minimum deterrence system in which States retained some nuclear weapons as a last resort, this factor might well become so decisive that the "last resort" could be quite safely dispensed with. In many primitive societies, for example, the force of tribal opinion is so great and the fear of ostracism so strong that no physical means of enforcement of community norms—no tribal police force—is needed. The same could be true in the global society that is now beginning to emerge. What Thomas Jefferson called "a decent respect for the opinions of mankind," reinforced by some conventional arms in the hands of a democratically determined central authority and by an implicit threat of economic sanction, could indeed make nuclear weapons totally unnecessary.

NARROW NATIONALISM

Many tangible facts illustrate the pervasiveness of nationalism in the latter part of the Twentieth Century. When people are thinking on the international level, fear is ordinarily fear of a national adversary; pride is pride in a national self; economic welfare is wanted for one's nation; rationalization is rationalization of one's nation's sins; projection is projection of blame onto a national adversary; community approval means approval by fellow nationals; and so on. In a sense, nationalism is the chief religion of the human race.

How, then, are we to evaluate nationalism? Is it mostly good or mostly bad? To what extent does it need to be overcome?

On the bad side, one is almost tempted to say that nationalism (defined here as a narrow identification much more with a particular

nation than with all of the other human beings on our tiny
endangered planet) is the primary, all-inclusive cause of war. If it
could be replaced by a "religion" of humanism, defined as a genuine
identification with the human race as a whole more than with any
nation, I have little doubt that the problem of preventing war would
be immensely simplified. The analogy of a lifeboat on a stormy sea
is appropriate. The nations are like individuals in such a lifeboat;
they will survive together or not at all. There will be common
security or no security. To recognize that fact is not altruism; it is
the grimmest kind of realism.

Yet there is a strong case for nationalism, which those who are
preoccupied with issues of war and peace often forget. It is now the
chief cement that holds a society together and permits the welfare of
the national whole to outweigh, in some contexts, the welfare of the
many individuals and groups contained within it.   History   and
everyday behavior alike testify to how readily human beings identify
with almost any in-group in contrast with almost any out-group.
Families, clans, tribes, gangs, corporations, religions, and races – all
can become the focus of an individual's prides and fears, as the
individual's self-image expands to become some kind of group self-
image. The process is facilitated when there is territory – "turf" –
to be defended against real or imagined threats from outside.
Robert Ardrey makes a strong case for the importance of territorial
defense in the human species as well as in many non-human animal
species.[3]

Such malleability in the nature of the group with which an
individual identifies and to which he or she feels loyal is encouraging;
it casts doubt on the dogmatic assumption that human beings are
incapable of graduating from a national to a world identification and
loyalty. Why not? Why couldn't they?

Only one good answer to this question has been suggested. It
is that human beings have a need not only to identify with one larger
group but also to have a common enemy that they can hate and fear
together – and feel superior to. There would be no such common
enemy – at least no such common human enemy, or inferior out-
group – if an individual genuinely identified with, and felt loyal to,
the entire human race. An all-too-human pride thus seems to
conflict with the psychological requirements for peace.

However, there are at least two good reasons to think that this
conflict may not be especially problematic. First, natural disasters
often have brought people together with great effectiveness, even
when no human agent was present to be hated, feared, or looked

down upon. Being adrift in a lifeboat on a stormy sea is such a situation too; the stormy sea functions psychologically as a common enemy might. The inevitability of mutual assured destruction (MAD) in an all-out nuclear war might also function psychologically as a natural disaster. In addition, unreconstructed nationalists might function for a long time as a common human enemy in the minds of the humanists who put their human loyalty above their national loyalty. The nationalists would not deserve to be hated because their nationalism would be so natural and, for them, inevitable; but they might be realistically feared and looked down upon as narrow-minded, simple-minded remnants of a bygone past that others had outgrown.

Returning to the concept of nationalism: Is it mostly good or mostly bad? Fortunately, the English language has three words that cover fairly adequately the three factors that need to be considered to answer this question: *patriotism*, which represents all that may be seen as good in the concept, *i.e.*, loyalty, self-sacrifice for the good of others, and so on; *chauvinism*, which represents all that may be seen as bad in the concept, *i.e.*, collective arrogance, power-seeking, "paranoid" thinking, and nationalism as a cause of war; and the word *nationalism* itself, which represents whatever good or bad may be associated with an identification with a particular nation. The nationalism of any nation or individual may involve, in other words, varying degrees of patriotism and chauvinism as defined.

For those of us who are primarily interested in peace, this approach suggests that we can wholeheartedly oppose the war-promoting characteristics of chauvinism, including the jealous clinging to national sovereignty that prevents effective organization of nations in larger units, while at the same time seeking to preserve the society-uniting characteristics of patriotism, even within such larger units. The World Federalists' definition of their aim as "World federation with minimum powers sufficient to prevent war" suggests a similar respect for the values of national patriotism.

EXAGGERATED FEAR OF BEING ATTACKED: A MAJOR CAUSE OF WAR

When conventional thinkers talk about the motives underlying war, they are likely to use words such as hate, greed, and craving for power. They are not likely to mention fear. Yet the historical record of acts of aggression in the Twentieth Century suggests that exaggerated fear—fear of being attacked by an opponent conceived as diabolically aggressive—is likely to be one of the major driving forces, even on the side of the chief aggressor.

The tendency to ignore fear as a cause of aggression, and therefore of war, is related to a simplistic Good Guys/Bad Guys conception of war itself. The conventional image of war portrays Bad Guys—motivated only by such disreputable impulses as hate, greed, and craving for power—engaging in cold-blooded, unprovoked attacks on Good Guys, such as one's own country or its allies, whose only motive is self-defense. The idea of a conflict between countries composed essentially of reasonably good guys, like oneself, and similar reasonably good guys on the other side, with some real guilt and much misperception on each side, is foreign to the thinking of most people.

Unfortunately, the aggressor who comes first to most people's minds when they think of war at all is Hitler, who *was* a Bad Guy and who had a relatively calculated long-term plan of aggression. What is not usually realized is that Hitler and Mussolini were to a surprising extent aberrations among Twentieth Century aggressors in the simplicity of their power motivation and in the appropriateness of the simple Bad Guy image when applied to them. The more typical pattern in this century has been quite different. Even Japan's motivation in World War II contained, apparently, a genuine though decidedly "paranoid" component of defensive motivation. Most of the Japanese appear really to have believed that they were in danger of economic strangulation by Western "imperialism"—with hostile powers encircling them on all sides (the United States, Communist Russia, and an inherently hostile, anarchic China)—unless they could control China and establish a Greater East Asia Co-Prosperity Sphere.

There are, of course, other examples, only a limited sampling of which is possible here:

- the German-Austrian attack on Serbia in 1914, probably motivated mainly by the fear that, if a "firm stand" against aggressive Serbian nationalism were not taken, Austria-Hungary would disintegrate and Russia would pick up the pieces;[4]
- the World War I German attack on France through Belgium, probably motivated mainly by fear of an imminent Russian invasion in the East and a two-front war;[5]

- the harsh treatment of Germany between 1918 and 1930 (which fueled Hitler's rise to power), motivated mainly by French fears of a third German invasion;
- though admittedly controversial, the Soviet takeover of Eastern Europe during 1944-1948, motivated largely by the trauma of World War II and a need for buffer territory to ease the fear of another invasion from the capitalist West, comparable to Hitler's onslaught in 1941-1945;[6]
- the 1948 Arab attack on the fledgling State of Israel, clearly motivated by the UN partition plan, which, in Arab eyes, amounted to a bare-faced seizure of Arab land, supported by imperialism, against which they had every right to defend themselves;[7]
- Israel's attack on Egypt in 1956 in collaboration with Britain and France, clearly motivated by Israel's perennial (and realistic) fear of being attacked and overwhelmed by the Arabs, who outnumbered the Israelis more than thirty to one and whose *fedayeen* raids seemed to demonstrate implacable hostility;[8]
- Israel's preemptive attack on Egypt and then Syria in 1967, largely motivated by the same perennial fear, much heightened by the fact that the Arabs seemed then to be on the verge of a concerted attack;[9]
- the Soviet crackdowns on Hungary and Czechoslovakia in 1956 and 1968, probably motivated in large part by the fear of a domino-style defection from the socialist camp if the unrest in either country had been allowed to go farther;
- the US intervention in Vietnam (whether or not called "aggression"), motivated primarily by the fear of a domino-style defection to communism throughout Southeast Asia and therefore a need to "stop communism" and, coextensively, preserve the credibility and ability of the US to deter communist aggression everywhere;[10]; and
- the Soviet invasion of Afghanistan (1979-1989), unequivocally an act of aggression and motivated primarily for the same or similar defensive geopolitical reasons (however exaggerated) that motivated the United States in Vietnam — in each case a superpower intervening by force in a small Third World country in support of a "friendly" but unpopular government threatened by a formidable popular revolt that in no way presented an imminent danger to the superpower.[11]

There are of course other instances in which defensive motives have been alleged by the aggressor but which are implausible enough to raise doubt as to their genuineness. Hitler's many aggressions, those of Japan, and that of North Korea in 1950 come to mind. However, such cases seem to be in the minority during most of this century (since 1913).

The generalization that many acts of aggression are motivated by fear raises a fundamental question as to how wars can be prevented. The conventional wisdom is that the most realistic strategy is armed deterrence—proving, through actual military strength and repeated evidence of resolve to use it when necessary, that a potential aggressor will have to pay an unacceptable price if aggression is dared. This is often said to be "the lesson of Munich," which refers to the fact that the failure at Munich to prove to Hitler that the West was willing to fight to stop his career of aggression was soon followed by further aggression on Hitler's part against what was left of Czechoslovakia and then against Poland.

The evidence is indeed strong that all of Hitler's later adversaries, including the United States and the Soviet Union along with Czechoslovakia (which had an excellent army and fortifications), would have been wise to stand united against Hitler at the time of the Munich conference. That probably would not have prevented a big war, as later evidence indicates that he actually wanted a war at that time; but it probably would have limited substantially the enormous war that did occur. And if the same kind of firmness had been shown earlier, before Hitler's actions in the Rhineland, in Spain, and in Austria, the great war might well have not occurred at all. However, no other national leader in this century has been a determined aggressor like Hitler. And Mikhail Gorbachev certainly is no such leader.

The paradoxical, counterintuitive generalization that many acts of aggression are motivated by fear is supported also by Ned Lebow's extraordinary book, *Between Peace and War*.[12] In it he describes thirteen "brinkmanship crises" during the past century and concludes that nearly always they involved one or both of two types of fear in the minds of those who challenged the status quo (*i.e.*, those who were in this sense the aggressors): fear of a *future* adverse balance of power in case they did not act decisively in the present; and fear of being badly hurt politically, at home, if they did not act decisively in the present. His prime example of the first of these (fear of a future imbalance of power) is Germany in 1914. Like many others, he attributes the German-Austrian aggression against Serbia primarily to fear that, if Serbia were not punished for the murder of the Archduke, Austria-Hungary would disintegrate and war would come at a later time when, with Russia meanwhile gaining in strength, Russia and France would have a decisive advantage over what was left of Germany and Austria-Hungary.[13] Other examples include the Berlin crisis of 1948, the Cuban missile crisis of 1962, and the Arab-Israeli crisis of 1967.[14]

Again, the basic important question – How can wars be prevented? – arises, and again the conventional wisdom is that the most realistic strategy is armed deterrence and threats that demonstrate resolve, proving to the Bad Guys that they will lose much if they dare to aggress. But armed deterrence and threats of physical harm carry with them a great risk: They *increase* fear in the opponent. And if aggression is caused mainly or often by too much rather than too little fear, increasing an opponent's fear can mean precipitating rather than deterring aggression. Lebow gives several examples. One is the Russian mobilization during the 1914 crisis. Intended partly to deter the Germans from attacking Serbia, it immediately heightened German fear that Russia and France together were about to start a big war and that in order to defeat its aggressive enemies Germany had to act preemptively – which it did.

The arms race itself has the same ambiguous character. Each step upward in an arms-race spiral is intended by its initiator to deter aggression by the other side. Yet the same step upward may be interpreted by the other side as further proof that the initiator has aggressive intentions and that more intensive arming on its part is therefore imperative.

The dilemma is clear: Do armed deterrence and threats that increase fear in the adversary increase or decrease the danger of war? In the Munich crisis and prior instances of Western appeasement the case is strong for thinking that greater strength and firmer threats might have prevented World War II. In the crises just cited the case is strong for thinking that they increased the danger of war. Which case is stronger? Or, better, under *what particular circumstances* do arms and threats tend to prevent war and under what circumstances do they tend to provoke it?

In *Perception and Misperception in International Politics*,[15] a seminal study, Robert Jervis suggests an answer: It depends, he contends, on the intentions of the adversary. If the adversary is a Hitler, determined to expand whenever the price of doing so does not seem too high, raising the price is a rational way to prevent aggression and war. But if both sides are in a spiral of mutual fear and mutual arming for the sake of deterrence, increased arms and threats are likely to have the opposite effect.[16]

Applying this idea to the Soviet-Western conflict since the death of Stalin (whose "paranoid" thinking was somewhat like Hitler's), the critical question is: What have the Soviet leaders' intentions been? Have their actions in East Europe, Afghanistan, and elsewhere been, like those of Hitler, part of a grandiose aggressive plan? Or have they been, like the somewhat comparable actions of the United States during the same period, primarily parts of an intermittent spiral in which each side's attempts to deter have been interpreted by the other as signs of Hitler-like aggressive intentions?

Whatever the answer to this question with regard to the long period from 1953 (when Stalin died) to about 1988, there seems little doubt that now, in the Gorbachev era, the Soviet Union does not have Hitlerite or Stalinist intentions and that therefore the primary emphasis of US policy should not be on arms and threats. Instead, strategic policies on both sides should be consciously geared to reducing, on the other side, at least the fear of being attacked. That is the kind of fear that promotes strenuous, defensively motivated arming and threats intended as deterrence in a time of crisis. Of course, the case for deterrence still is strong in the case of would-be aggressors like Adolph Hitler.

What seems needed, in sum, is a combined strategy of *deterrence and reassurance* on both sides. This conclusion is so crucial that there is a need to explore its psychological aspects at some length.

REASSURANCE: MORE IMPORTANT THAN DETERRENCE

The upshot of the previous discussion is that there is too much fear in the world. And it is the wrong kind of fear. Too often, instead of a healthy, realistic fear of the resistance that would be encountered if one nation attacks another (which is the kind of fear that all deterrers hope to create in their opponents' minds), it is an unhealthy, greatly exaggerated, "paranoid" fear of being attacked by a diabolical, inherently aggressive enemy that does not actually exist. In their one-eyed concentration on every form of national military strength, nations increase the danger of a catastrophic war, and cripple themselves economically in the process.

There is an answer and, though not simple, it is clear. In the present-day world, and in a future post-nuclear world, deterrence should be limited to what actually is needed, and greater emphasis should be placed on a systematic strategy of reassurance – relieving a potential opponent's fear that we are diabolically aggressive, relieving our own fear that the opponent is diabolically aggressive, and strengthening the positive bonds that actively promote peace.

The first question, then, is: How much armed deterrence and what kinds of armed deterrence actually are needed?

There is little doubt that some is needed now. Some peace activists go so far as to question the need for any armed deterrence, but history does not support that extreme position. The lesson of Munich is real and should be heeded, even though neither Gorbachev nor his potential successors are Hitlers. The work of Jervis[17] and Lebow[18] makes it plain that deterrence sometimes works and sometimes is counterproductive. The question of the circumstances in which it works and those in which it does not is very complex and is by no means fully answered as yet, but educated guesses can be made now, as is indicated in the previous discussion.

In any event, there are several Twentieth Century instances in which armed deterrence seems to have worked well, at least in the short run.

- From 1918 to 1938, Germany, intensely humiliated by the "Diktat" of Versailles, was very hostile to France and aspired to restore its former hegemony in Central Europe. Before Hitler came to power, the Germans were far too weak to attempt any such risky adventures and even after his arrival the predominant power of France and its allies prevented him, until 1938, from taking the serious risk he then took. Even Hitler, with his paranoid delusions of grandeur and his quite exceptional relishing of war as such, was effectively deterred from 1933 to 1938.
- From 1945 until the late 1980s, the Soviet Union and the United States, though intensely hostile toward each other during most of that period, engaged in no actual fighting against each other. Whether their fear of nuclear war as such was the chief reason is a controversial question, but the existence of that much hostility during that long a period without actual fighting is presumptive evidence that nuclear deterrence worked at least in part.[19]
- Specifically, during the 1956 Hungarian uprising and the 1968 "Prague Spring" many Americans believed that it was their obligation to risk war to help the Hungarians and

Czechs free themselves from Soviet domination. The United States did not do so. No doubt, the chief reasons were fear of war as such, plus fear of probable defeat on the "home territory" of the Soviet army, which had the advantage of the defensive role.

- Most recently, the United States appears to have been similarly deterred by the strength of the Soviet army in Afghanistan and by the logistical difficulties of that inaccessible area—though sympathy with the Afghan "freedom fighters" was very strong and was matched by military assistance.

Reviewing these examples of seemingly effective deterrence, we can make an educated guess that armed deterrence is likely to work (1) when it takes the form of adequate arms—threats and demonstrations of "resolve" being another matter; and (2) when the arms are not excessive relative to what is sufficient to prevent the undesired behavior on the part of the opposition. Thus, an *emphasis on deterrence* (in conjunction with strategies of reassurance) probably could include both of the following minimal options without endangering world peace.

1. Enough conventional strength to deter the other side without having to rely on nuclear weapons. There is ample reason to believe, particularly as of this writing when all of Eastern Europe is being reconstituted along more democratic lines, that a policy of conventional deterrence on the part of the West as well as the East would not require very much in the way of military hardware and expenditure. Precisely how much is enough or how much is too much is of course a question that is best left answered by specialists knowledgeable in military and economic affairs. The formula "enough conventional strength to deter an invasion" probably should take into account, however, the precarious and unpredictable circumstances of present-day Eastern Europe.

2. As a stop-gap measure in today's anarchic world, enough *second-strike* nuclear capability to deter the other side from even considering the initiation of nuclear war. Following Robert McNamara's estimate that one-twentieth of the West's current nuclear strength would be adequate for this purpose,[20] this means that only a very few nuclear-armed submarines and bombers would be needed to deter potential nuclear enemies.[21] Of course, such an option would not be available in a world truly free of nuclear weapons—that is, in a world in which one could trust and verify the

total or near-total absence of nuclear weapons altogether. But recognizing that nuclear weapons can be "reinvented" and that such reinvention could take place in secret, minimum deterrence would seem wise. Indeed, in an intermediate period, minimum deterrence policies of some sort might even be essential.

The preferred greater *emphasis on reassurance* (in conjunction with the previously-noted strategies of deterrence) should include all of the following in a post-nuclear world and, to the extent possible, in the present-day world.

1.   Total elimination of all *offensive* weapons of mass destruction (including chemical and biological weapons), by agreement if possible but unilaterally if necessary. In my judgment, nothing short of an explicit threat of their use can increase an opponent's fear of being attacked as much as does the mere possession of offensive weapons of mass destruction.   Their elimination is therefore a number-one priority for a strategy of reassurance.

2. Scrupulous avoidance of explicit or implicit nuclear threats (comparable to the Strategic Air Command [SAC] alert of 1973) in times of crisis. A quiet possession of arms ("speaking softly while carrying a big stick") is more justified by the historical record than either explicit or implicit threats of actually using them. In Ned Lebow's terms, "general deterrence" is less dangerous than "immediate deterrence."[22]

3. Active pursuit of confidence-building measures (CBMs), which allows potential adversaries to reduce the possibility of conflict caused by incorrect assessments of the other's military movements. Standard examples would include notification of military exercises and movements above a defined level, limitation of military maneuvers to low tactical levels close to national frontiers, prohibitions on live ammunition during military exercises and movements, and exchanges of observer personnel and liaison.

4. Stress the essentially irrational nature of the image of the diabolical enemy, through education and the mass media, so that both policy-makers and ordinary citizens in one's own country will not be inclined to approve unrealistic policies premised on that image.

This last recommendation is among the most important that can be made. Increasingly in recent years, even before *glasnost* and *perestroika*, concerned Western scholars and other professionals have sought to look more objectively at, for example, the Soviet Union to enhance their comprehension of that country's culture, history, security, and values so as to at least minimize, if not altogether avoid, the potential for misunderstanding and conflict.[23]   Urie Bronfenbrenner, a US psychologist who could speak fluent Russian, led the way for psychologists by reporting what he heard in many conversations with ordinary US and Soviet citizens in a landmark article published in 1961.[24] According to Bronfenbrenner, five major themes predominated, among them the following two.[25]

They Are Aggressors

The American View: Russia is the warmonger bent on imposing its system on the rest of the world. Witness Czechoslovakia, Berlin, Hungary, and now Cuba and the Congo. The Soviet Union consistently blocks Western proposals for disarmament by refusing necessary inspection controls.

The Soviet View: America is the warmonger bent on imposing its power on the rest of the world, and on the Soviet Union itself. Witness American intervention in 1918; Western encirclement after World War II with American troops on every border of the USSR (West Germany, Norway, Turkey, Korea, Japan); intransigence over proposals to make Berlin a free city; intervention in Korea, Taiwan, Lebanon, Guatemala, Cuba. America has repeatedly rejected Soviet disarmament proposals while demanding the right to inspect within Soviet territory–finally attempting to take the right by force through deep penetration of Soviet airspace (the U-2).

Their Government Exploits and Deludes the People

The American View: Convinced Communists, who form but a small proportion of Russia's population, control the government and exploit the society and its resources in their own interest. To justify their power and their expansionist policies they have to perpetuate a war atmosphere and a fear of American aggression. Russian elections are a travesty, since only one party appears on the ballot. The Russian people are kept from knowing the truth through a controlled radio and press. . . .

The Soviet View: A capitalistic-militaristic clique controls the American government, the nation's economic resources, and its media of communication. The group exploits the society and its resources. It is in their economic and political interest to maintain a war atmosphere and engage in militaristic expansion. Voting in America is a farce, since candidates for both parties are selected by the same powerful interests, leaving nothing to choose between.

Bronfenbrenner continues in the same vein, giving many examples under the three remaining major headings: "The Mass of Their People Are Not Really Sympathetic to the Regime," "They Cannot Be Trusted," and "Their Policy Verges on Madness."

Testimony of a similar sort (but presented by a hardheaded, skeptical newspaperman instead of a psychologist and given twenty years later in 1981) is Robert Kaiser's description of the Soviet people's response to their government's invasion of Afghanistan:[26]

> When the Soviets realized that they had caused an international furor much stronger than they had expected, they reacted bitterly, if also typically. Dr. Freud could have had the Russian nation in mind when he devised his theory of projection. How could this be our fault? Clearly the Americans were to blame for the unhappy change in the international atmosphere brought on by the Soviet Union's fraternal help for the people of Afghanistan. This was the Party line, and because it suited the national personality, a great many Russians obviously accepted it. Like people everywhere the Russians are gifted at presuming their own benevolence.

Several points in this remarkable passage deserve attention: the important role of propaganda ("the Party line"), the similarly important role of special receptivities in citizens' minds ("it suited the national personality"), the obvious connection between incipient guilt feelings about what their government was doing in Afghanistan and a need to blame others (*i.e.*, the Freudian mechanism of projection), the simultaneous process of rationalization (they had only given "fraternal help" to the "people" of Afghanistan), and the recognition that rationalization and projection are not exclusively Russian mechanisms ("like people everywhere").

There are, of course, beyond the quest for better mutual understanding, further actions and non-actions that adversaries, including the superpowers, can consider in implementing a policy that emphasizes reassurance more than deterrence. One is a scrupulous avoidance of actions in the Third World that might even look like aggression in the eyes of the other side. Another is strengthening the peacekeeping powers of the United Nations and strong financial support of the UN. And still another is full cooperation with the other superpower as well as the UN in regional matters such as Southern Africa and the Middle East. If the will is there, the ways will open.

PSYCHOLOGICAL ROOTS OF THE
EXAGGERATED ENEMY IMAGE

I described previously the kind of fear that needs to be reduced by a strategy of reassurance as "an unhealthy, exaggerated, 'paranoid' fear of being attacked by a diabolical, inherently aggressive enemy that does not actually exist." The kind of enemy image that leads to that kind of fear is, in my judgment, the chief direct psychological cause of war. It fuels arms races, leads to defensively motivated aggression, and, in a post-nuclear world, even could predispose a decision-maker to "reinvent" nuclear weapons for preemption purposes. If that fear is to be reduced, we would be wise to try to understand it—and its roots. It can be illustrated in extreme form in the words and actions of three individuals who have been perhaps the most outstanding "war-causers" in our century: the Kaiser Wilhelm in 1914, Adolph Hitler, and Joseph Stalin.

When the German Kaiser discovered that the little war that he wanted to fight against Serbia was likely to be transformed into a great European war and that the great British Empire was likely to be ranked among his enemies along with Russia and France, he wrote these words on the margin of a diplomatic note:

> The net has been suddenly thrown over our head, and England sneeringly reaps the most brilliant success of her persistently prosecuted, purely *anti-German world policy* against which we have proved ourselves helpless, while she twists the noose of our political and economic destruction out of our fidelity to Austria, while we squirm *isolated* in the net.[27]

If we define paranoia as delusions of persecution plus delusions of grandeur, usually with the delusions of persecution more prominent, then this passage can be fairly called paranoid, even if we say about the Kaiser (and about Hitler and Stalin) that they were not psychotic but paranoid "within the normal range."

The Kaiser's outburst should be seen in its immediate historical context. Lord Grey, the British Foreign Minister, had done his level best (not necessarily wisely) to prevent a world war, and the Kaiser, along with his chancellor, Theobold von Bethmann-Hollweg, had been pursuing a policy of outright war against the independence of a small nearby country, Serbia. During this historical period there was guilt aplenty on both sides. The British Empire in previous years had much to answer for, but at this particular time the genuine guilt was primarily on the German and Austrian side. Yet the

Kaiser saw only total innocence on his side ("our fidelity to Austria")
and total guilt on the side of Britain ("sneeringly twists the noose").
It is a classical case of the Freudian mechanism of projection, the
essence of which can be represented by the formula: We are not to
blame, *they* are.   The collective self-image is preserved from a
realistic sense of guilt by projecting all guilt onto the image of the
enemy.

The flavor is similar in the following quotation from Hitler,
and it, too, is a classic example of projection:

> Behind every murder stood the same power which is responsible for this
> murder; behind these harmless, insignificant fellow-countrymen who
> were instigated and incited to crime stands the hate-filled power of our
> Jewish foe, a foe to whom we had done no harm, but who none the less
> sought to subjugate our German people and make of it its slave – the
> foe who is responsible for all the misfortune that fell upon us in 1918,
> for all the misfortune which plagued Germany in the years that
> followed.[28]

And strikingly similar is a quotation, in Nikita Khrushchev's
memoirs, from Joseph Stalin: "You'll see; when I am gone the
imperialistic powers will wring your necks like chickens."[29]
Khrushchev's memoirs go on to explain:

> He [Stalin] lived in terror of an enemy attack. For him foreign policy
> meant keeping the aircraft units around Moscow on a twenty-four hour
> alert. . . . He was also notoriously suspicious of his colleagues. . . .
> Stalin was a very distrustful man, sickly suspicious. We knew this from
> our work with him. He could look at a man and say: "Why are your
> eyes so shifty today?" or "Why are you turning so much today and
> avoiding to look me directly in the eyes?"[30]

It is something of a commonplace in psychology to say that a
line between the "abnormal" and the "normal" is hard to draw and
that even the most abnormal individuals merely carry to an extreme
certain potentials or tendencies that exist in all of us. The tendency
to Good Guys/Bad Guys thinking that Bronfenbrenner found to be
so startlingly similar in the thinking of ordinary US and Soviet
citizens is a case in point.[31] It can be argued that the delusions of
persecution and of grandeur that seem so bizarre in a Kaiser
Wilhelm, a Hitler, or a Stalin carry to a near-psychotic extreme the
familiar processes of denial and rationalization on the Good Guy
side and of projection on the Bad Guy side that tend to characterize

us all when we find ourselves caught in situations of acute group conflict.

Summarizing the reasons for the diabolical enemy image, what lessons may we derive for enhanced security in a post-nuclear world? Consider the following reasons:

1. *Reality.* There are always or nearly always kernels of truth even in the most paranoid national delusions. Therefore, in the present-day world as well as in any future world, there are and probably will continue to be real injustices that need to be redressed by peaceful democratic means. There also are real enemies, actual and potential, that may have to be deterred by force. Accordingly, armed deterrence sufficient to make the other side think long and hard before committing further acts of aggression is, unfortunately, a matter of simple prudence.

2. *The self-interest of many opinion-molders.* Both in the Soviet Union and in the United States, there is a political-military-industrial complex which has acknowledged and unacknowledged reasons of power-maintenance and privilege-retention for deliberately or half-deliberately sustaining the diabolical enemy image.

3. *The receptivity of the general public.* Everywhere or nearly everywhere there seems to be a special receptivity to the Good Guys/Bad Guys caricature, with the mechanisms of denial and rationalization working overtime to sustain the Good Guys image and the mechanism of projection working overtime to sustain the Bad Guy image.

It follows that three main elements in a strategy emphasizing reassurance, as an alternative to the present overemphasis on deterrence, should be:

1. to give others no basis in reality for maintaining their diabolical image of one's own country or of a regional or world authority;

2. to diminish the influence of groups gaining from international conflict in one's own country, including, for example, a policy of economic conversion to reduce the self-interest in sustaining an arms race;[32] and

3. to educate people in one's own country regarding the nature of the mechanisms of denial, rationalization, and

projection in the international context and to note objectively the specific facts that contradict the Good Guys/Bad Guys picture of any particular conflict—for example, by appropriate teaching of history and current affairs—so that realistic empathy with "the other fellow's point of view" can be encouraged in many ways.

"MACHO PRIDE" AND "AGGRESSION"

"Macho pride" is not an altogether familiar term. It is defined here as an *excessive* desire for power, prestige, and a virile self-image, although the word "excessive" is likely to raise eyebrows among those who, like this writer, normally have a strong preference for operational, non-evaluative definitions that make it easy to judge whether a particular thing fits a definition or not. That is not true of this definition. Here, excessiveness is the essence of what the definition is intended to convey, and it is necessarily, emphatically evaluative. In fact, the proposition to be defended here is that, though a normal, healthy amount of concern with power, prestige, and virility cannot be said to promote war, an excessive amount of it on the international level does greatly—and tragically—promote war. It has contributed much in our century (and more so in previous centuries) to the following kinds of behavior:

1. *Empire-building.* Hitler in Europe and Japan in China are outstanding examples.
2. *Empire-keeping.* Austria-Hungary's attempt to put down nationalist strivings in its ramshackle empire was a major initial cause of World War I. What if Austria had been content to accept the modest but self-respecting style of life that it was forced to accept after 1918? Would World War I have occurred? What if the Soviet Union were now fully prepared to accept the boundaries suggested by the extent of Great Russian, Ukrainian, and Byelorussian speech (they are very broad)? Would not the danger of major international conflict be greatly diminished?
3. *Arms races.* It has sometimes been suggested that the fantastic amount of overkill on both sides of the superpower nuclear arms race is not really due to a need for security, on the Soviet side, as much as it is due to the fact that the USSR needs its nuclear strength to claim to be

a superpower at all. Might not the same ego-need exist in the United States as well?

4. *Forcibly taking disputed territory.* When there is overlapping of territorial claims by two peoples (*e.g.*, the Polish and German claims to the Polish Corridor in 1939 or the Israeli and Palestinian claims to the West Bank) and when one side unilaterally and forcibly tries to settle the issue, there is nearly always trouble and sometimes war.[33]

5. *Humiliation-avoidance*: During the earlier part of the crisis of late July 1914, Austria, France, Germany, and Russia each seemed to be primarily concerned with the danger that, if it did not "stand firm," it would "sink to the status of a second-class power." It was only after July 29 that fear of imminent war and of defeat in war took over from macho pride as the chief apparent motive on each side. Only then was the macho "game of chicken" replaced by the military men's need to preempt. In the Cuban missile crisis of 1962 the United States seriously risked nuclear war. As President John F. Kennedy put it, he needed choices other than "humiliation or surrender"; and the resulting humiliation of the Soviet Union marked the beginning of its especially dedicated pursuit of the nuclear arms race.

It can be argued, too, that macho pride, rather than genuine fear or economic motives, at least partly explains the frequent use of force by the United States in the Caribbean area. Many US citizens, including our national leaders, appear to have felt it intolerable to see those "banana republics" and "little dictators" thumbing their noses at the great United States. But because the United States has safely dominated the area militarily since 1898, it is hard to believe that exaggerated fear—even fear of an increased Communist foothold in the area—actually has been the chief reason for the US use of force.

Are there psychological alternatives to macho pride? Can it be significantly diminished so as to enhance global security?

In this century, there are three striking examples of macho pride being significantly diminished. Germany changed after World War II, perhaps largely because of the combination of complete, unequivocal defeat and fairly magnanimous treatment by the victors. Japan changed similarly, probably for the same reasons. And, after many years of vainly trying to match or surpass the whole of the West in economic and military strength, the Soviet Union appears now to have begun to relax and to resign itself to a status of one of the great powers but not necessarily one of two equal superpowers. Reason itself was not primarily responsible in any of these three cases. It required the shock of defeat or long-continued failure, plus a certain minimum of reasonableness on the part of opponents. Might it be otherwise? If not, the augury is not good for curing the United States of its macho pride. Thankfully, military defeat does not appear to be in the offing. Is the United States destined for prolonged failure – environmental, economic, and so on?

ANGER AND "AGGRESSION"

Is "aggression" an instinct, innate in the human species? For lack of space, this much-debated question is bypassed here; but two fairly obvious and related propositions are worth noting:

1. Whatever the role of innate factors, wars are *not* inevitable. They depend on circumstance. The ninety-nine-year period of comparative peace in Europe from 1815 to 1914 is good evidence that this is true.
2. Whatever the role of innate factors, anger and hate (which can be defined as hardened, long-term anger) are important contributors to violence and war. Witness, for example, Northern Ireland, the Israelis and Palestinians, Ayatollah Khomeini's feelings about "the Great Satan," and the quotations from Kaiser Wilhelm, Adolph Hitler, and Joseph Stalin (*see* this chapter: Psychological Roots of the Exaggerated Enemy Image). Anger and hate tend to occur whenever one nation perceives another, realistically or not, as seriously threatening its security, its territory, its economic well-being, its pride.

Therefore, *unnecessary* anger or hate in opponents' minds, caused by unnecessary threats to security, territory, economic well-being, or pride, should be scrupulously avoided by those who want peace. Only through realistic empathy can such threats – and wars that may result from them – be avoided.

ECONOMIC IRRATIONALITY

The world has spent an estimated $16 *trillion* dollars for military purposes since World War II.[34] The United States alone is now budgeting about $300 *billion* annually for military purposes,[35] not counting the hidden expenses of the brain-drain from the civilian industry into the war industry, which saps the productivity of the US economy in its competition with others such as the Japanese and the Germans. The Japanese, it should not be forgotten, spend only about 1 percent of their GNP on preparation for war.

Economic analyst Ruth Sivard had a field day describing the other purposes that could be served if military spending were not such a debilitating drain on the world's economies.[36] "The world's annual military budget," she wrote, "equals the annual income of 2.6 billion people in the 44 poorest nations."[37] She noted also that "[t]he world spends 2,900 times as much on national military force as on international peacekeeping forces"[38] and that "[o]ne aircraft carrier (Nimitz class) costs $3,900,000,000, which equals one solid meal a day for 6 months for the 20 million Americans who do not get enough to eat."[39]

These costs fade into insignificance, however, when compared to the economic costs of any big war. Even a relatively tiny war such as the war in Vietnam, which the Johnson Administration did not dare to charge *directly* to the taxpayers, was costly enough to set the United States on the borrowed money binge that continues to this day. Indeed, it should not be forgotten that a drastic reduction of the annual deficit of the US Government currently is being made impossible in large part because of the $300 billion annual bill the United States pays for its armed forces—at a time when, with the breakup of the Soviet empire, the Soviet danger looks smaller than it has for the past forty-odd years. Fiscal responsibility, since President Lyndon Johnson's deceptive way of paying for the Vietnam conflict, has been another casualty of the arms race.

Considering how basic economic motives are in human affairs generally,[40] the fact that these expenditures are accepted so complacently by so many presents an enormous paradox. Why haven't economic motives been more powerfully mobilized to do the things, including drastic diminution of the arms race, that for other reasons seem necessary for world peace?

A frequent answer is that although the general taxpaying public suffers economically, those who profit from the arms industry gain, and their political power is greater than that of the majority who suffer. For those who stress this factor, the "military-industrial complex" (known in the Soviet Union as the "capitalist ruling circles") is the crux of the problem.

This answer may explain a large part of the problem, but it leaves unanswered another key question: Why does the general voting public—which enormously outnumbers the military-industrial complex, and which includes a great many well-to-do taxpayers who do not, on balance, profit from the arms industry—supinely accept the tax burden and the other losses that the arms industry entails? This is a psychological question, and psychologists should be at least looking for answers to it.

As a psychologist, I suggest two answers. One is the familiar fact that any special interest, such as that of the military-industrial complex, is likely to be fully aware of its interest and well organized in pursuit of it, while the general public is likely to be much less aware and much less organized. The other answer is that, in the minds of the public, there are special reasons for selective inattention to the costs of a great arsenal.

Let us say that a US citizen or member of Congress is considering the military budget. He or she may be hardly thinking in budgetary terms at all. Those who present the issue probably are framing it in such terms as "Do we need a strong defense?" or "Don't we live in a dangerous world, and in such a world don't we need a strong defense?"—the implied question being "Which is more important, having more creature comforts or survival itself?" In addition, an implicit appeal to a strong and virile self-image seems fairly clear. Former Secretary of Defense Caspar Weinberger often spoke to the US public in this way, and when the issues are thus posed only one kind of answer seems properly patriotic, realistic, and toughminded: "Of course the world is dangerous, and of course survival is more important, and of course we Americans are tough enough and realistic enough to choose survival."

But complex and inconvenient questions then are ignored, much to the disadvantage of those who fail to ask them:

> "Will our buildup cause the Soviets to build up even more?"
> "Aren't we already, basically, the strongest and safest nation in the world?"
> "Doesn't the arms race increase the danger of war?"
> "Hasn't the Soviet danger gone down considerably since Gorbachev came in?"
> "What else will we have to go without (such as paying our debts or feeding 20 million hungry Americans) if we continue the arms race at its present level?"

Like the others, the economic question, in spite of the general salience of economic well-being in voters' minds, apparently tends to get lost once a decisive answer is given in favor of a policy that seems to offer, simultaneously, national security, a tough and virile self-image, and a properly patriotic attitude. Rather than face the value trade-offs that have to be candidly recognized if the conflict between these things and the economic motives is allowed to become clearly conscious, the individual avoids the trade-offs by not following the train of thought into all of the skeptical questions that, realistically, ought to be asked.

OTHER PSYCHOLOGICAL FACTORS

Overconfidence[41]

Geoffrey Blainey, an Australian historian, has written a non-quantitative but extremely interesting book entitled, simply, *The Causes of War*.[42] Based on his study of some two hundred wars since 1700 A.D., Blainey concludes that the chief common factor causing them was overconfidence.

In the Twentieth Century there have been many examples of wars, overconfidently started, that ended in defeat (or stalemate after an exhausting war): Germany and Austria in 1914, Hitler when he attacked the Soviet Union, Japan in China, Japan at Pearl Harbor, North Korea when it invaded South Korea, the United States when it invaded North Korea, the Arabs when they attacked the infant State of Israel, Israel when it invaded Lebanon, the United States when it intervened in Vietnam, Iraq when it attacked Iran, the United States in Cuba and Nicaragua, the Soviet Union in Afghanistan. Of course, such a summary leaves out many factors. But two clear reasons for overconfidence stand out: an initial failure to anticipate accurately the fierce, tenacious resistance that the victims of aggression exhibit, and a failure to anticipate the intervention of others on behalf of the victims. A psychological factor underlying most of these failures of anticipation stands out: an absence of realistic empathy with the victims of aggression, and with the other nations that intervened on behalf of the victims.

Absence of Empathy: A Form of Selective Inattention

Empathy (defined here simply as understanding how others perceive their world – not as agreeing with them or as feeling sorry for them) is the great antidote to the misperceptions that cause war. It operates in at least two major ways: by counteracting the diabolical enemy image (which, as we have seen, underlies nuclear overkill, the nuclear arms race, and defensively motivated aggression) and by counteracting the overconfidence that has led many nations to commit aggression without realizing the price they would eventually have to pay for doing so.

An absence of realistic empathy is a form of selective inattention, defined here as an unconsciously motivated turning of attention away from evidence that conflicts with unconsciously cherished images. An image of one's own country as morally innocent requires a fairly obvious, generally recognized form of selective inattention (and self-deception). It obviously underlies defense mechanisms such as denial, rationalization, and projection. What is less often recognized is the "need for an enemy" that makes the diabolical enemy image similarly cherished.[43] And unconsciously motivated selective inattention—to the understandable fears in the minds of the "enemy" people, to the understandable anger in "enemy" minds at what one's own nation has done or is doing, and to the intense desire for peace that usually exists in "enemy" minds—makes cultivation of realistic empathy the urgently needed antidote that it is.

There are many ways of cultivating realistic empathy with opponents, allies, and neutrals: teaching history with emphasis on empathy with both sides in every conflict; role-playing; travel; citizen exchange; reading or listening to authentic expressions of the viewpoints of all sides; conflict-resolution workshops; experiential training in conflict resolution that includes genuine, attentive listening to the other side's point of view. The lasting security of a world that is free of nuclear weapons depends on all of these activities.

CONCLUSION

Many people now approve of the greatly strengthened regional and world organizations envisioned here, but they despair at ever achieving them. They are seen as unrealistic, visionary, utopian in the worst sense. Is such despair now realistic—if ever it was? Is it realistic in the light of the startling changes of the past few years, especially the general movement toward democracy, the great domestic changes within the USSR, and the breakup of the Soviet empire? Might a world that is largely or wholly free of nuclear weapons be achievable in as little as, let us say, thirty years?

For persons lacking a real sense of history, thirty years seems so far in the future that even thinking about it now appears as foolishly impractical and visionary. It would be at least 2020 A.D. before anything would be realized, and futurists who talk about a date in the third millennium, a date beginning with 2 instead of 1, always have sounded at least faintly ridiculous. But for those with a real sense of history, a time perspective limited to less than thirty years is ridiculously truncated and shortsighted. (Did any "realist" honestly anticipate the reconstitution of Eastern Europe that now is taking place a year or even six months before the process began?) Within thirty years, barring major catastrophe, many children now of primary school age will be in the prime of life and many adults now in the prime of life still will be alive. It is a perspective-promoting exercise to think back to thirty years ago — no longer ago than 1960 as of this writing. Do you remember what you were doing in 1960? Parents and educators today are having an influence on the thinking of students who may be presidents, prime ministers, or members of Congress, and quite probably will be voters, a mere thirty years from now. Think also about some of the substantial changes most of us have seen during the past thirty years: the emergence of the civil rights movement, the change in gender relationships and sex mores, the coming of computers, the changes in China's economic system, the rush of change recently in the Soviet Union and Eastern Europe. Or think of the development of the European Community over just a few short decades.

It is not, in other words, impractical or visionary or too soon to think concretely about what kind of world authority would be desirable and about what might be the most practical ways to achieve it. All unfamiliar destinations, especially preferred ones, require road maps.

It goes without saying that a narrow nationalism, insisting on a kind of sovereignty that regards "being outvoted by foreigners" as the greatest of all evils, is an enormous obstacle to the attainment of such a world, and would continue to be an obstacle even if there were much more effective organizations on the international plane. But in this respect a psychological distinction of great importance must be made: the distinction between the defensive spirit of national independence that resists imperialistic domination by other countries (a defensive spirit that is more prominent in our century than in any previous century) and the offensive spirit of national aggrandizement that seeks to dominate other countries (which was thoroughly discredited, though not altogether eliminated, by the outcome of World War II). The religion of nationalism that pervades our era is not the Hitler type of nationalism. It is the spirit of legitimate national self-determination in non-military affairs, as opposed to and opposing intrusive foreign control.

It follows that regional and world organizations with minimum powers sufficient to prevent war—or domination of any nation by any other nation—might actually be welcomed by most of the nations in the world if they saw it in that light. The problem becomes, then, the problem of getting them to see it in that light; and to this end there must be credible guarantees of the right of every nation, subject to human rights obligations, to do what it likes within its own borders, including the right to determine its own form of government and economic system. Precisely how this might be accomplished is largely for the international lawyers and political scientists, not the psychologists, to determine; but its psychological desirability and feasibility are within the province of social and political psychology, and I, as a political psychologist, think it could be done in as little as thirty years if those who believe in it regain the hopes that many of them had in 1945 and 1946. Politics is not only the art of the possible. It is also the art of realistic hoping.

Some important other psychological trends, in addition to the growing conviction that nuclear war must be prevented, are working in favor of peace and should be recognized. There are the technological and economic changes that are making our planet smaller and smaller by promoting awareness of common needs and by giving experience in international cooperation: new techniques in transportation and communication, expansion of trade, expansion of free trade areas, multinational corporations (and the need to control them by democratic means), a common investment market, and the need for a common currency. There are other needs also, neither economic nor political, that call insistently for effective cooperation: control of drug traffic, control of terrorism, cooperative preservation of the environment and a livable climate, the non-proliferation of nuclear raw materials, and so forth.

It is encouraging to note also that the factors that will promote world organization in the long run are desirable also for peace in the short run. Thinking and talking about strengthening the United Nations, for instance, including its peacekeeping forces and peacemaking powers, will immediately increase the world's ability to surmount international crises, while also increasing the long-term prospect of achieving a world without nuclear weapons. The same can be said about an arms-control regime that would limit nuclear weapons to a minimal, invulnerable nuclear deterrent on each side of the East-West conflict—perhaps one-twentieth of what now exists.

Whatever increases the ability of people to work together in coping with short-run dangers also will contribute to the psychological foundations of a sane long-run organization for peace. There is no contradiction between the short run and the long run.

Finally, there is great basis for hope in the steadily growing solidarity among at least the continental countries of the European Community, and, at this writing, the year 1992 gives promise of becoming a landmark in that development. With its nucleus consisting of close cooperation between France and West Germany and with growing evidence that it represents economic gain for its members, the European Community seems likely to become the vital attracting center of a potential world organization, probably drawing in the increasingly independent East European countries and perhaps eventually also the United States, Japan, and even the Soviet Union. The area of solid peace described by Bruce Russett in this volume can, to paraphrase the words of the "Internationale," become the human race.

NOTES

1. *See* this chapter: Narrow Nationalism and Conclusion.

2. *See* this chapter: Exaggerated Fear and Reassurance.

3. *See* Robert Ardrey, *The Territorial Imperative* (New York: Dell, 1966).

4. *See* Sidney B. Fay, *The Origins of the Great War* 2 (New York: The Free Press, 1966): 552; Hajo Holborn, *A History of Modern Germany, 1840-1945* (New York: Knopf, 1969): 415.

5. *See* Ole Holsti, Robert North, and Richard Brody, "Perception and Action in the 1914 Crisis," in David J. Singer (ed.), *Quantitative International Politics* (New York: The Free Press, 1968): 136-38; Barbara Tuchman, *The Guns of August* (New York: Bantam, 1976): 75, 94.

6. *See* Marshall Shulman, *Stalin's Foreign Policy Reappraised* (New York: Atheneum, 1965): 20, 26.

7. Ralph K. White, "Misperception in the Arab-Israeli Conflict," *Journal of Social Issues* 33, 1 (1977): 198-99, 206-08.

8. Howard M. Sachar, *A History of Israel* (New York: Knopf, 1976): 429-90. There probably were more realistic ways to cope with the danger, but the fear was real.

9. Sachar, *History of Israel*: 615-38; White, "Misperception": 209-10.

10. Neil Sheehan *et al.* (eds.), *The Pentagon Papers* (New York: Bantam, 1971): xix, 205-12; Ralph K. White, *Nobody Wanted War: Misperception in Vietnam and Other Wars* (New York: Doubleday/Anchor, rev. ed. 1970): 182-240.

11. *See* Raymond L. Garthoff, *Détente and Confrontation: American-Soviet Relations from Nixon to Reagan* (Washington, DC: The Brookings Institution, 1985): 887-965; George B. Kennan, "Imprudent Response to the Afghanistan Crisis?" *Bulletin of the Atomic Scientists* (April 1980): 7; Strobe Talbott and Bruce Nelan, "The View from Red Square," *Time* (February 4, 1980): 18.

12. Richard Ned Lebow, *Between Peace and War: The Nature of International Crisis* (Baltimore: Johns Hopkins Press, 1981).

13. *Ibid.*: 334.

14. *Ibid.*: 59.

15. Robert Jervis, *Perception and Misperception in International Politics* (Princeton, NJ: Princeton University Press, 1976).

16. *Ibid.*: 58-113.

17. *Ibid.*

18. Lebow, *Between Peace and War*. *See also* Richard Ned Lebow, "Conventional vs. Nuclear Deterrence: Are the Lessons Transferrable?" in *Journal of Social Issues* 43, 4 (1987).

19. *See* Lebow, "Conventional vs. Nuclear Deterrence": 184.

20. *See* Robert S. McNamara, *Blundering into Disaster: Surviving the First Century of the Nuclear Age* (New York: Pantheon, 1986).

21. Implicit, of course, is the elimination of all nuclear overkill capability. To see an opponent armed with nuclear weapons far beyond what it needs for deterring a nuclear attack inevitably suggests that its weapons are not intended merely for deterrence purposes, that indeed they may be intended either for intimidation or attack. Robert McNamara's recommendation to reduce nuclear weapons perhaps to one-twentieth of their present level makes sense. *See* McNamara, *Blundering Into Disaster*. It could be done in stages, as in Charles Osgood's strategy of Graduated and Reciprocated Initiatives in Tension-reduction (GRIT), but the faster the better. *See* Charles E. Osgood, *An Alternative to War or Surrender* (Urbana, IL: University of Illinois Press, 1961), *reprinted in part in* Burns H. Weston (ed.), *Toward Nuclear Disarmament and Global Security: A Search for Alternatives* (Boulder, CO: Westview, 1984): 337-44.

22. Lebow, "Conventional vs. Nuclear Deterrence": 183-84; Richard Ned Lebow and Janice Stein, "Beyond Deterrence," *Journal of Social Issues* 43, 4 (1987): 89.

23. *See generally* Daniel N. Nelson and Roger B. Anderson (eds.), *Soviet-American Relations: Understanding Differences, Avoiding Conflicts* (Wilmington, DE: SR Books, 1988), including a chapter therein by Burns H. Weston, "Who Are the Soviets? The Importance of Accurate Perception in the Age of Trident." *See also* Burns H. Weston (ed.), *Toward Nuclear Disarmament*: ch. 6 ("Rethinking 'The Enemy'").

24. Urie Bronfenbrenner, "The Mirror Image in Soviet-American Relations," *Journal of Social Issues* 17, 3 (1961).

25. *Ibid.*: 74-75.

26. Robert Kaiser, "U.S.-Soviet Relations: Goodbye to Détente," *Foreign Affairs* 59, 3 (1980): 511.

27. *Quoted in* Robert C. North, "Perception and Action in the 1914 Crisis," *Journal of International Affairs* 1 (1967): 115.

28. Adolph Hitler, as quoted in Raoul de Russy de Sales, *My New Order* (New York: Reynal and Hitchcock, 1941): 345.

29. Nikita Khrushchev, *Khrushchev Remembers*, Strobe Talbott trans. and ed. (Boston: Little, Brown, 1970): 392.

30. *Ibid.*: 393, 585.

31. *See* text at and following note 25.

32. For pertinent discussion, see the chapter by Lloyd J. Dumas in this volume.

33. There are many other examples: Alsace-Lorraine, the Sudetenland, Quemoy, Kashmir, the Sino-Indian border, the Ogaden region in Ethiopia, and so on.

34. *See* Michael Renner, "Enhancing Global Security," in Lester Brown (ed.), *State of the World, 1989* (New York: Norton, 1989): 133.

35. *See* William W. Kaufman, "A Defense Agenda for Fiscal Years 1990-1994," in John D. Steinbrunner (ed.), *Restructuring American Foreign Policy* (Washington, DC: The Brookings Institution, 1989): 63.

36. *See, e.g.,* Ruth Leger Sivard, *World Military and Social Expenditures, 1987-1988* (Washington, DC: World Priorities, Inc., 1987).

37. *Ibid.*: 5.

38. *Ibid.*

39. *Ibid.*: 35.

40. *See, e.g.,* Hadley Cantril, *The Pattern of Human Concerns* (New Brunswick, NJ: Rutgers University Press, 1965).

41. A condensation of my recent paper, "Why Aggressors Lose," *Political Psychology* 11, 1 (1990).

42. Geoffrey Blainey, *The Causes of War* (New York: The Free Press, 1971).

43. In this connection, see the quotation from Georgi Arbatov at the outset of the chapter by Bruce Russett in this volume.

7

Religion and Alternative Security: A Prophetic Vision

William Sloane Coffin

Imagine the vanity of thinking that your enemy can do you more damage than your enmity.

—St. Augustine

According to a prophetic vision widely shared by many religions, we all belong one to another, everyone of us on this planet. That is the way God made us. From a Christian point of view, Christ died to keep us that way, which means that our sin is only and always that we are putting asunder what God has joined together. The answer to the ancient question, "Am I my brother's keeper?" is really, "No, Cain, you are your brother's brother." In other words, human unity (including sisters as well as brothers) is not something we are called upon to create so much as it is something we are called upon to recognize—and to make manifest.

All the great religions speak this essential truth, as does patriotism at its best. True patriots are no more uncritical lovers of their country than loveless critics of it. Mindful that territorial discrimination can be as evil as racial or religious discrimination, religious patriots carry on a lover's quarrel with their country as a reflection of God's eternal lover's quarrel with the whole world. As the great Spanish cellist Pablo Casals is reputed to have said: "The love of one's country is a splendid thing, but why should love stop at the border?"[1]

THE PROPHETIC VISION
AND THE CONTEMPORARY WORLD

No doubt this vision of human unity will strike many as hopelessly idealistic. In fact, however, it becomes desperately realistic the moment we realize that it is the world as a whole that needs now to be managed, and not just its parts. Fifty years ago it was enough to be concerned with the parts; mainly, one part of the world had to protect itself against another part. Today it is the whole that cannot protect itself against the parts.

Nowhere is this new reality more evident than in the realm of war and peace. In World War II, nations at war targeted one another. Today, all nations live on the target of World War III. The whole world is as a prisoner in a cell, condemned to death and awaiting the uncertain moment of execution. Taking human unity seriously is thus a first and essential point for lasting world peace, for survival. A world clamoring to be free of nuclear weapons and aspiring to be rid of war itself simply cannot do without human unity.

The same prophetic vision also holds that nature is not separated from nature's God, that it is filled with wonder and sanctity, and that all human beings are spiritually linked to every creature and leaf—an understanding that surely will appear naive, not to say quaint, to the descendants of René Descartes, who accurately predicted that we would become "the masters and possessors of nature." But how masterful is our mastery? The radioactive fallout that arose from the meltdown at Chernobyl recognized no frontiers. Neither do toxic wastes and acid rain. The devastation of our oceans and forests affects all biological species everywhere. So does damage to the ozone layer. Global warming is a present concern as annually our cars emit into the air their weight in carbon. And so once again we are forced to recognize that it is the world as a whole—its inhabitants *and* their environment— that has to be managed, and according to policies that will cease treating nature as a toolbox and extend to it some of the same ethical considerations that inform our personal relations.

Additionally, the ancient prophetic vision teaches us that peace is not merely the absence of conflict but, as well, the presence of justice.[2] Said Amos: "Let justice roll down like mighty waters." He did not say: "Let charity roll down . . ."—and for a reason well discerned by none other than Karl Marx, who once observed to a gathering of church people: "You Christians have a vested interest in unjust structures, which produce victims to whom you then can pour out your hearts in charity." It is not that charity is not important. It is, in fact, so important that everyone should have a part in it. But charity is no substitute for justice. If peace is the presence of justice, the world will have ultimately little peace either if the prosperity of a minority is obtained at the price of the permanent poverty of the majority or if the prosperity of the majority is obtained at the price of the permanent poverty of a significant minority. When the economic oligarchs self-righteously give "alms for the poor" without ever questioning the social and economic order that makes their charity necessary in the first place, they help to create, literally, revolting conditions. To quote Amos again: "You think to defer the day of misfortune but you hasten the reign of violence." Or as President Kennedy correctly observed, "[t]hose who make peaceful revolution impossible will make violent revolution inevitable."[3]   And in our increasingly interdependent world, revolutionary parts have a nasty way of threatening the otherwise tranquil whole.

A final point to underscore about the ancient prophetic vision is that it took the future seriously, never forgetting its goal. It pictured God *ahead* of us as much as above or within us. "And it shall come to pass in the latter days . . ." are words that always are followed by an image of the world at one and at peace. This point is important because we will not have a future if we do not think about the future. Recognizing that thinking about the future requires imagination, we need to imagine a world preferable to the predictable one. We need to imagine a world whose citizens will be as mindful of international law as they are of domestic law and so obey the decisions of the World Court and other international tribunals. We need to imagine a world whose peacekeeping forces will be larger than any national force; a world whose agencies will be supported by an international income tax based—why not?—on energy consumption. We need to imagine a world not free of conflict, for the horizons of the world always will be darkened by dissension, but a world free at least of violent conflict, of deadly weapons, of toxic wastes, of abject poverty, and of other demonstrations of structural violence. And if all of this, and especially the prospect of a warless world or a world free of nuclear weapons, seems again like vapid idealism, that may simply indicate how far we have slipped behind a schedule we should have kept in the first place—if, that is, we are serious about saving our planet.

Obviously this prophetic vision of human unity—this vision of peace and ecological balance and social justice—will be understood only by those people who go to Sabbath worship not to escape the world but to find a better one. It will be understood, and hopefully followed, by people who see religion as *this*-worldly as well as other-worldly. Although it is true that many Christians say "If you believe in Jesus you can live forever," it would be more Biblical of them to say "If you believe in Jesus you can live well." Christianity is more life-affirming than death-denying, and in today's world the truly religious question is less "What must I do to be saved?" than "What must we all do to save one another?" Believers, after all, are stewards of God's creation, which now is at risk in an order of magnitude never previously even imagined.

A WORLD LIVING BEYOND ITS MEANS

In 1948, General of the Army Omar Bradley commented that "[o]urs is a world of knowledge without wisdom, of power without conscience. We know more about killing than we know about living, more about war than we know about peace."[4] And of the United States he said, "[w]e are a nation of nuclear giants and ethical infants."[5]

Bradley's words reflect what we might call a biblical anthropology, an understanding that humanity is not perfectible. Rather, humanity seems permanently poised between "truth and endless error." When pressed, people almost always prefer their own interests to those of their neighbors. And thus, on top of our predilection for folly and perversity, it really should come as no surprise that humanity "makes a poorer performance of government than of almost any other human activity."[6] Lord Acton was correct in observing that "power tends to corrupt, and absolute power corrupts absolutely."[7] He could have added that, in its own way, powerlessness also tends to corrupt and absolute powerlessness corrupts absolutely.

Humankind's imperfectibility thus being what it is, it would have been interesting to have asked General Bradley if he could have conceived of the United States, or any other nation for that matter, ever becoming an ethical giant; it would have been interesting to ask him if he could have conceived of nations ever becoming practicing disciples of the prophetic vision. In all probability, of course, Bradley would have demurred, as would any person who knows that nations have even less chance of being perfect than do individuals. Just as they live beyond their financial means, so also do nations live beyond their ethical means.

Actually the United States and the Soviet Union did both in the 1980s; they went into extraordinary debt by pursuing an arms race that gambled recklessly with the fate of humanity. It is, in fact, little short of miraculous that the world has so far been spared a nuclear weapons exchange or accident, for which it is no more prepared than was the Soviet Union for the meltdown at Chernobyl or the United States for Exxon's ecological disaster in Prince William Sound — which is to say, of course, that nations tend to live also beyond their rational means.

In theory, nuclear deterrence makes sense. It makes sense that in the nuclear age nations should fear not losing a war but fighting one in the first place. But as nuclear deterrence is far from foolproof, it is in practice a wicked thing. And to pretend to be in control when you are not is criminal folly. Wrote British historian Herbert Butterfield: "The hardest strokes of heaven fall upon those who imagine they can control things in a sovereign manner as though they were kings of the earth playing Providence not only for themselves but for the far future . . . , gambling on a lot of risky calculations in which there must never be a single mistake."[8]

Thus, nations with nuclear weapons are "living in sin" (to use an old pietistic phrase). The morality of extrication may be complicated, but the mere existence of nuclear weapons is morally intolerable. Only God has the *authority* to end life on this planet; we human beings have only the *power*. Until we destroy every last nuclear weapon on earth, and for that matter all other weapons of mass destruction, we shall continue to live beyond humanity's ethical and rational means.

But even if we succeed in ridding the earth of nuclear weapons and other weapons of mass destruction (including chemical and biological weapons), the ability to make them — to "reinvent" them — will be part of the storehouse of human knowledge forever. Were two nations to engage one another in conventional warfare and were one of them to begin to lose, it is would be utterly naive to assume that the losing party would go gracefully down to defeat rather than resort to nuclear weapons had it the knowledge to make them or otherwise gain access to them.

In short, it is not enough to say that warfare relying on weapons of mass destruction (nuclear and otherwise) is today pathologically dysfunctional. It is not enough for nations to insist on intrusive verification, to challenge inspections without right of refusal. It is not enough for nations to negotiate treaties and other international arrangements by which nations can reach a consensus for the control, reduction, and elimination of all weapons of mass destruction. Such pronouncements and initiatives are of course important, for the world never will be safe until people genuinely comprehend the pathological ways of their national leaders, until all laboratories are open for inspection, until all the nations consider themselves bound by the international law of peace. But if genuine long-term security ever is to be achieved, we have to stare hard and long at the fundamental truth that humanity has outlived war and doesn't know it.

In other words, only by coming to grips with this reality will humankind be able to fulfill the prophetic vision of human unity, to evolve a new planetary ethos of human solidarity. That is not to say that we must look, in dreamy-eyed fashion, to end all greed and destructive conflict. We need not ask the impossible. But surely we can and must expect a world in which human beings will cease being their own worst enemies because they have reached a new level of problem-solving and ethical awareness appropriate to the age of interdependence in which they live. They will have to recognize that God is not mocked, that we have to be merciful when we live at each other's mercy, that we have to be meek or there will be no earth to inherit.

WE'RE ALL IN THE SAME BOAT

Why is it so hard for people to realize that what Abraham Lincoln once said of his country now applies to the whole world: "A house divided against itself cannot stand?"[9] Why can we not see that villages become towns, towns become cities, cities become nations, and now nations must become a global village?

One reason is fear—fear of the unknown, which makes us averse to change. Often human beings resemble the proverbial caterpillar who, looking up at the butterfly, said: "You'll never find me flying around in one of those crazy things." Yet change is as insistent as death and taxes, and if God is ahead of us then we are only faithful as long as we remain pilgrim people.

Another reason is pride, which is not accidentally but essentially competitive. In church history we see how the faithful need infidels to confirm them in their fidelity. Likewise nations need other nations to prove to themselves and others that they are *more* powerful, *more* prosperous and—God help us!—more virtuous. Perhaps more than anything else, it is belief in their moral superiority that makes nations so intransigent. It blinds them, for example, to the kinds of double standards we saw again and again in the 1980s. The Soviets armed; it was evil. The United States did the same; it was necessary for national security. Soviet satellite interceptors were "killer satellites;" US interceptors went by the acronym "SAINTS." We condemned the Soviet invasion of Afghanistan, while we continued to believe that in Central America, as earlier in Vietnam, the United States had the right to decide who lives and dies and rules in other lands. The Soviets were that "evil empire;" the United States was the "leader of the free world." "Hypocrisy," wrote François de la Rochefoucauld, "is the homage which vice renders to virtue."[10]

But the peoples of the world never will become one in love until they first admit that they are all one in sin. It is important to recognize that all nations make decisions based on self-interest and then defend them in the name of morality. It is important to recognize that a nation's salvation, like that of an individual, lies not in being sinless but in believing that there is more mercy in God than sin in us. It is God who is too good to damn us, not we who are too good to be damned.

When I stated that "God is not mocked," I meant that we do not so much break the Ten Commandments as we are broken on them. The world seems to swing on an ethical hinge: Fuss with that hinge, and eventually history and nature feel the shock. Racial discrimination, a nuclear arms race, the thoughtless devastation of nature—something immoral eventually proves dumb. Even as I write, the clear winners of the Soviet-US arms race are Japan and West Germany! There also are fewer songbirds as they find less of the forest that is their home.

All of this underscores the importance of enlightened self-interest. Religion and morality will play a part, but it is essentially enlightened self-interest that will lead the nations of the world to moderate national sovereignty and increase global loyalty. It is enlightened self-interest that will lead them to abandon national security for global security. For if we are all in the same boat, as indeed we are, it makes no sense to bore a hole in the other fellow's side. Every successful reform movement in history has combined the promise of great material benefits for great numbers of people with clarity of moral purpose. To see this is to see how self-interest and ethical considerations work well together.

Self-interest and ethical considerations work especially well in relation to nuclear disarmament and, more comprehensively, complete and general disarmament. For it would appear that the eminently ethical business of dismantling our nuclear and conventional arsenals is inextricably linked to our consummate self-interest in ecological stability and socioeconomic justice—without which there simply cannot be any *real* security. Only by serving the cause of disarmament can sufficient funds be found to service the requirements of a sustainable ecosystem and a just social and economic order. Of course, there is no guarantee that the funds liberated would be used to serve these crucial ends. And this is a fact devoutly to be rued. But as true believers in the prophetic vision we must accept, as Eleanor Roosevelt taught us, that "it is better to light a candle than to curse the darkness."[11]

STEWARDS OF THE EARTH

Consider, thus, our global environment. I said earlier that we need to extend to nature some of the same ethical considerations that presently govern our personal relations. We need a "land ethic," although I am not optimistic about our ability to devise one. Our desecration of the environment will not necessarily prompt us to renew the spiritual ties that once bound us to nature. On the contrary, if we do not recognize nature's God, our impulse may well be to adapt not ourselves to nature's ways but nature to ours, to master nature even more than we already have done.

Enter biotechnology, for example. In *The End of Nature*, Bill McKibben suggests that genetic engineering may prove the most significant invention since the discovery of fire.[12] For where we have always worked with what was given, now we can create new forms of life. If nature cannot coexist with our numbers and habits, then we will simply create a new nature—that is, engineer plants and trees and crops and animals that can survive in any environment we choose to create.

To do all this, concededly, even to think it, demands perhaps more hubris than heroism. Certainly it could be said to put us in the deity business, telling God who's in charge.

On the other hand, to give hubris its due, it is well to remember that Buckminster Fuller did not see himself as an enemy of the environment, but rather as a champion of the human race. He compared our present evolutionary state to that of a chick in a shell. In our natural shell, we have been nourished for years with coal and oil and other forms of energy. "But then," as Fuller noted, "by design, the nutriment is exhausted just at the time the chick is large enough to be able to locomote on its own legs. And so, as the chick pecks at the shell seeking more nutriment, it inadvertently breaks open the shell."[13]

To Fuller, as to other enthusiasts of the new age to come, it is right for human beings to break out of the age-old confines of nature. After all, we are the brains of the earth, the planet managers. What we presently are facing is only our final evolutionary examination. Of course, it is anything but certain that we will pass this examination. We could easily flunk, because in warring with nature there are real risks of losing in winning. But that we plan to take the exam seems pretty much a foregone conclusion.

The alternative is to insist that if industrial civilization is killing nature as we know it, then we must radically transform industrial civilization as we know it. The German philosopher Martin Heidegger once compared windmills and dams. The former, he observed, do not affect the wind though the latter clearly affect rivers and all that surrounds them. Analogously, we should compare our dependence on fossil fuels with the endlessly renewable energy we could harness from the sun, the wind, and the tides. This requires, of course, that we see the end of hubris as nemesis and its alternative as humility.

That is to say, following Heidegger, that we must choose to be faithful stewards of our fragile planet rather than its managers. Wrote John Muir: "How narrow we selfish, conceited creatures are in our sympathies! How blind to the rights of all of the rest of creation."[14] He would have us recognize what Chief Seattle and so many other Native Americans always have known, that we are but one species among many, that nature belongs not only to us but to God and to all the wild things that call it home. McKibben sees it correctly, I think; we face a profound spiritual choice: to remain God's creatures or to make ourselves gods. Genuine global security, which inescapably insists upon ecological sanity, depends on our remaining God's creatures.

THE NEED FOR AN EQUITABLE
DISTRIBUTION OF WEALTH

I also noted earlier that the world will never know true peace — what Kenneth Boulding has called "stable peace"[15] — if the prosperity of a minority is obtained at the price of the permanent poverty of the majority or if the prosperity of the majority is obtained at the price of the permanent poverty of a significant minority. It is an alarming prospect and a likely one too, because the prosperous of the world tend to fear the poor more than they pity them. Thus, prosperous nations are generally more concerned with disorder than they are with injustice, and the almost invariable result is to produce more of both, as we have seen recently south of both the Soviet and US borders.

Scholars have long puzzled over the meaning of the phrase in the book of Genesis: "and God hardened Pharaoh's heart." They should, for it is difficult to conceive of God's love being anything less than human love as we know it at its best. But whatever the phrase may mean, it would seem to confirm that justice will be obtained through the organizing efforts of the poor rather than through the belated decency of the rich.

History, though not predetermined, will not dispute this conclusion. Also, it will demonstrate that revolutions do not produce new human beings, only other human beings. As every revolution produces a new ruling class, then people who are deeply concerned that justice "roll down like mighty waters" must become, as it were, permanent revolutionaries. With wealth and power continually at stake, battles fought and won in the arena of peace and justice rarely stay won.

Concern for peace and justice, however, must encompass the rich as well as the poor. In the Bible it is always the rich who are a problem to the poor, never the other way around. But the biblical view is that judgment *of* the rich spells mercy not only for the poor but, ultimately, for the rich as well. Just as the poor should not be left at the mercy of their poverty, so the rich should not be left at the mercy of their wealth.

Proof perfect of this contention was the decade of the 1980s. In the United States, and elsewhere I suspect, it was an odious decade. "Enrich thyself" was the clarion call. Our cities saw the deterioration of everything not connected with profit-making. Hunger and homelessness haunted the land. An observant author called our Congress "the best that money can buy."[16] The 1980s brought to mind what William James wrote to H.G. Wells at the turn of the century: "[T]he moral flabbiness born of the exclusive worship of the bitch goddess success – that, with the squalid cash interpretation put on the word success, is our national disease."[17] The 1980s were a showcase for the scriptural question: "What shall it profit a man if he gain the whole world and lose his soul?"

The truly renewable resources are intellectual, aesthetic, spiritual—not economic. And though it is true that some immensely wealthy people brilliantly renew their spiritual as well as their other resources, the virtue of the few hardly takes care of the inertia of the many. A second-century manual on church discipline addressed well-to-do Christians in this manner: "If you are willing to share what is everlasting, you should be that much more willing to share things which do not last." A more equitable distribution of the world's wealth would represent not only greater justice for the poor but, as well, a measure of salvation for the rich. It is, in any event, an absolute condition precedent for world peace.

CONCLUSION

There is, of course, an intensely personal dimension to all of this. It is in our personal lives that religion and morality make an enormous difference. The world never will have true peace as long as we cling to our own. Nor will righteousness flourish as long as we keep asking "What can I get out of life?" instead of, "What can life get out of me?" Political transformation is heavily dependent on personal transformation.

Said Eleanor Roosevelt: "[I]t isn't enough to talk about peace. One must believe in it. And it isn't enough to believe in it. One must work at it."[18] Oddly enough, those who must work hardest are probably those outside of government. I say this because power blinds before it corrupts, so much so that those farthest from the seats of power tend to be nearer the heart of things. Certainly that has been the case in all the great reform movements of the United States. Moreover, in a democracy such as in the United States, we should not bow to Washington as do Moslems to Mecca; we should look to Washington only to ratify what it can no longer resist.

We citizens—in the United States and elsewhere—must become the agents for the change we want to see in the world. God put us humans on earth to make it a better place and God can help us if only we will avail ourselves of the divine aid that always is at hand. It is not that we must prove something to ourselves or to the world. Truly religious people know that God's love has taken care of that; they know that it is not because we have value that we are loved, but because we are loved that we have value. Our value is a gift, not an achievement. And with nothing to prove, we are free to express ourselves. Certain it is that true peacemakers are not those who have something to prove, only those who have much to express.

Peacemakers are those who make a gift of themselves—which means no guilt, no self-hatred. For how do we make a gift of that about which we feel guilt, of that which we hate? All of us have much for which we can feel guilty, but forgiveness is precisely for that which we cannot condone. Again, there is more mercy in God than sin in us.

To fight national self-righteousness without personal self-righteousness is hard but necessary. Lasting world peace depends on it. Never should we fight evil as if it were something that arose totally outside ourselves. As the saying goes, it takes a sinner to catch a sinner! A good spiritual exercise is to find something to love in those we hate and to find in ourselves that which is unacceptable in others.

Of course, it is pure sentimentality not to realize that the implementation of compassion demands confrontation. Yet like Gandhi and Martin Luther King, we must contend against wrong without becoming wrongly contentious. And when disillusionment sets in, as inevitably it will, we must be particularly severe with ourselves. We must ask ourselves: "Who in the world ever gave me the right to have illusions in the first place?" Optimism is not necessary. What is needed is hope, whose opposite is not pessimism but despair. Like the sea around a small craft, despair cannot sink us until it gets inside. Frequently I have said to myself: "If Jesus never allowed his soul to be cornered into despair, who am I to succumb?"

Burnout, to be sure, is a threat to all. I doubt if there is a peacemaker alive who has not at some point felt like Rocinante, a tired hack of a horse ridden by a Quixote idea. But it helps to recall that ultimately in this world God calls upon us to be faithful, not effective. Thus we can be persistent, even when we cannot be optimistic. We can keep the faith despite the evidence, knowing that only in so doing has the evidence any chance of changing. And if we cannot change evil, at least we can fight to make sure evil does not change us.

It is of course when people are grasped by the power of love that we have the greatest chance of changing evil, the greatest chance of fulfilling the prophetic vision of human unity. For it is love that enhances our sense of belonging one to another—rejecting uniformity, insisting on mutuality. It is love that fosters engagement with the world, not retreat from it. People who love good even more than they hate evil, who love peace even more than they hate war—these are deeply religious people, people in whom God believes even though they may profess no belief in God. And deeply religious people, who seek no reward in life other than the integrity that comes with being in the right struggle, refuse to tolerate the intolerable. They believe that the present order, no matter how dark and dangerous, is never the last word. They are willing to defy the odds, convinced with St. Paul that if we fail in love we fail in all else. And they take comfort in the poet's assurance that we are "[o]nly undefeated because we have gone on trying."[19]

NOTES

1. The exact source of this quotation is unknown, but it has been attributed to Casals.

2. *See, e.g.*, Martin Luther King's "I Have A Dream" speech given at the Civil Rights March on Washington on August 28, 1963, in James Melvin Washington (ed.), *A Testament of Hope: The Essential Writings of Martin Luther King* (San Francisco: Harper and Row, 1986): 217.

3. From a March 12, 1962, White House address to Latin American diplomats. *Quoted in* Emily Morton Beck (ed.), *Familiar Quotations/John Bartlett* (Boston: Little, Brown, 1980): 891.

4. From a speech delivered December 10, 1948, in honor of Armistice Day. *Quoted in* Emily Morton Beck (ed.), *Familiar Quotations/John Bartlett* (Boston: Little, Brown, 1980): 825.

5. *Ibid.*

6. Barbara W. Tuchman, *The March of Folly: From Troy to Vietnam* (New York: Alfred A. Knopf, 1984): 4.

7. Letter to Bishop Mandell Creighton from Cannes, April 5, 1887, *reprinted in* J. Rufus Fears (ed.), *Selected Writings of Lord Acton: Essays in the Study and Writing of History* 2 (Indianapolis, IN: Liberty Classics, 1986): 378.

8. Herbert Butterfield, *The Origins of Modern Science* (London: Hazell, Watson and Viney, 1949).

9. "A House Divided" speech delivered at Springfield, Illinois, on June 16, 1858, *reprinted in The Collected Works of Abraham Lincoln* 2 (1848-58) (New Brunswick, NJ: Rutgers University Press, 1953): 461.

10. Duc François de La Rochefoucauld, *Reflections: Or Sentences and Moral Maxims* (New York: Dodge Press, 1900): 218.

11. Motto of The Christopher Society, which was a favorite of Mrs. Roosevelt and is widely associated with her.

12. *See* Bill McKibben, *The End of Nature* (New York: Random House, 1989): 154, 160-70.

13. *Quoted in ibid.*: 156.

14. *See generally* "Editor's Preface: John Muir and the Rights of Animals," in Lisa Mighetto (ed.), *Muir Among the Animals: The Wildlife Writings of John Muir* (San Francisco: Sierra Club Books, 1986).

15. *See* Kenneth E. Boulding, *Stable Peace* (Austin, TX: University of Texas Press, 1979).

16. Phillip Stern, *The Best Congress Money Can Buy* (New York: Pantheon, 1988).

17. Letter to H. G. Wells (September 11, 1906), in Henry James (ed.), *The Letters of William James* 1 (Boston: Atlantic Monthly Press, 1920): 253.

18. November 11, 1951, Voice of America broadcast.

19. T. S. Eliot, "The Dry Savages," in *Four Quartets* (New York: Harcourt, Brace, 1943): 21, 28.

8

Toward Post-Nuclear Global Security: An Overview

Robert C. Johansen

Where there is no vision, the people perish.
— Proverbs 29:18

The delegates who met in Philadelphia at the Continental Congress in 1776 proclaimed a Declaration of Independence because they believed their liberties could best be secured by complete separation from a powerful political rival, the British Crown. Later, to protect their newly won liberty and to maintain peace more generally, US officials followed what was then and remains today probably the most widely believed maxim of international relations: "If you want peace, prepare for war." As George Washington stated in the first annual presidential address to both houses of Congress: "To be prepared for war is one of the most effectual means of preserving peace."[1]

The world has grown more interdependent since 1776. Today it brims with military, economic, and environmental interpenetrations that crisscross all territorial boundaries. Under these new conditions, attempts to secure liberty through political separation and to maintain peace through preparation for war are bound to fail. Today, the interpenetration of societies means that people in the United States cannot enjoy liberty if people elsewhere threaten an attack against which there is no defense. Although the blessings of liberty remain as precious as they were in the late 1700s, the most

220

effectual means of preserving peace have changed fundamentally. No one's life, liberty, or pursuit of happiness can be realistically achieved through sovereign disengagement from other countries. Indeed, as the chapters in this volume vividly demonstrate, the peace and security of one society can be achieved only through the active enhancement of the peace and security of rival societies. To be secure, all States need reliable international agreements or laws that specify what others may or may not do with their military power, their hazardous wastes, and even their economic resources.

Moreover, today as in 1776, liberty and security can be encouraged by making government accountable to people. But unlike the Eighteenth and Nineteenth centuries, when the blessings of liberty and peace could flow from nations taking separate diplomatic paths, today *each* government must be held accountable to *all* people affected by its major decisions, whether they live within that government's borders or not. No society, through separation or adversarial relations, can escape duties to others or protect its rights against others. Integration, not separation, is the yet largely unspoken watchword of the hour, the underlying need for common understandings that is addressed by alternative security concepts, policies, and institutions.

The purpose of this chapter is to synthesize and move beyond the previous chapters in order to provide an overall orientation for meeting the new security needs of the United States and world publics. I begin with a brief discussion of the meaning of alternative security and then move to an examination of how, in light of the preceding chapters, the concept of alternative security can open the windows of the imagination to envisage an integrative, boundary-transcending set of principles to guide world security policy.

THE MEANING OF ALTERNATIVE SECURITY

The terms "common security" and "comprehensive security" encapsulate the two fundamental qualities that distinguish the new concept of security from the old. Security can be achieved today only when rival nations hold it in common and only when people take a comprehensive view of security threats that encompass demographic, economic, environmental, political, psychological, and religious as well as military problems that jeopardize their future. For security to be held in common means that the United States, among others, must give sincere attention to maintaining the security of its rivals as well as of its own public and allies. If in the future a powerful adversary like the Soviet Union or another nation feels threatened, it might well increase its ability to threaten the United States or indeed engage in combat that, even if not directed against the United States, would almost certainly inflict severe security costs on the United States from radioactive fallout, global economic dislocation, climate change, or irresistible migratory pressures from unwanted populations. In an environmentally fragile age where nuclear and non-nuclear weapons of mass destruction may be grasped at almost any moment by many different nations, the United States cannot maintain its own security while ignoring or even inadvertently increasing the insecurity of other societies.

Indeed, the old and still-prevailing goals of national security, if pursued with traditional policies and preparations for war, actually destroy the prospects for common security. Today's exponents of more sophisticated nuclear and conventional arms — those who endorse a $300 billion military budget for the United States and sincerely see themselves as the best of US patriots — unintentionally constitute a new kind of internal enemy, probably more dangerous to the blessings of liberty than was the British monarch of the late 1700s. They unknowingly threaten US security by standing in the way of common security policies that are the only hope for war prevention in the long run.

The comprehensive nature of security needs constitutes the second emphasis proposed here. It pushes the conceptual change even further by moving it from military security, which in the past has been attained through a haphazard balance of military power, to safety from all of life's major threats, which in the future can be addressed through a wide variety of new international regimes and institutions governing global environmental, economic, military, and human rights issues. To borrow some computer jargon, if in the past the militarily competitive balance of power system has been the default setting for interstate behavior in a relatively anarchic world, in an age of interpenetrating forces and military threats that cannot be repelled we need to utilize a more sophisticated program that can ensure that conflicts will be conducted through agreed-upon non-lethal procedures, that resource-sharing will proceed in an orderly fashion so the system will not crash, and that common values will be implemented efficiently through cooperative arrangements.

As a more comprehensive concept of security begins explicitly to incorporate non-military threats, the concept of "enemy" also changes. Those people who by their consumption patterns and driving habits intensify atmospheric carbon dioxide and thin the ozone could well constitute a more serious security threat to the United States than does a peasant movement demanding land reform in Central America or the unlikely threat of a nuclear attack from the Soviet Union. Yet many of the perpetrators of these largely unacknowledged, non-military security threats are found not in some distant land or in the embrace of a foreign ideology but, instead, among the governing elites and consumer segments of the United States and other friendly industrialized societies. Many of the contemporary Western security threats originate from within the Western camp, not from without. Thus, security proposals that dwell primarily on inter-adversarial diplomacy—the traditional focus—simply are not bold enough to address the full range of security problems.

ENVISIONING ALTERNATIVE SECURITY GOALS

Whereas prevailing security policies emphasize national sovereignty and military competition in the balance of power, exponents of alternative security envisage a growing ability to form bridging identities and institutions among nations. To be sure, new strategies of security enhancement must operate in the existing diplomatic framework, but they also must guard against merely perpetuating it, as traditional national security managers and inertial social forces are bound to demand. A powerful and effective future security policy must aim deliberately at establishing a new structure of international relations and a code of conduct that reflects the mutuality and comprehensiveness of security.

This aim can be sustained through the rough and tumble of politics with an intellectual map or animating vision of what must happen to enable the species to survive with dignity. That vision can be expressed conveniently as the ability to transcend five time-honored boundaries, both tangible and conceptual, that now exacerbate all contemporary dangers to human life by fracturing the human mind, spirit, and community, and that consequently impede movement toward common, comprehensive security.

The first of these boundaries is the species-dividing boundary between nation-states, a physical-psychological boundary that produces easy acceptance of the "us-them" approach to economic and political life and the Good Guys/Bad Guys syndrome that Ralph White so compellingly shows to be a recurring problem in international relations.[2] If this boundary and its accompanying "territorial discrimination"[3] are not transcended with a heightened sense of human solidarity, collective violence and globally undemocratic political decisions will accompany the human race for as long as we are able to put off extinction. Intra-species national identities can be—often are—carried to extremes when they are wedded to military power, so that people in one sub-species group begin to think of other people with a different group identity as less than human. This separatist identity, which Erik Erikson called "pseudo-speciation" three decades ago,[4] obviously facilitates killing and obstructs alternative security policies.

The second boundary is between rich and poor, both within and between nations, perpetuating economic inequity and widespread, unnecessary suffering and conflict. As Lloyd Dumas has demonstrated, gross economic inequities create the conditions that often give rise to violence and squander opportunities to employ economic relationships to build a strong structure of peace.[5] The right kind of economic relationship, not just the extent of economic connections, produces peace.[6] This point is illustrated by differences between the separatist, war-inducing function of French-German economic relations after World War I and the integrative, pacifying function of economic relations between France and Germany since World War II. Indeed, a balanced economic relationship not only helps keep the peace, it also is a more efficient economic relationship in the long run,[7] of enormous importance in the coming age of scarcity. Yet the failure to narrow the gap between rich and poor in the years since 1945 demonstrates how widely people have accepted this self-destructive boundary in practice if not in rhetoric.

Third, an intergenerational boundary, which seldom is bridged, enables people recklessly to consume the planet's resources and overload it with pollutants without concern for future generations (or indeed for anyone's long-term needs) and without respect for generations who have gone before us. Short-term considerations dominate political decisions, corporate balance sheets, military decisions about how to achieve security, and even moral calculations.

Fourth, boundaries between the human species and the rest of nature must be transcended to increase respect for all of creation and to stop despoiling the biosphere. The prospect of population growth, deforestation, desertification, ozone depletion, and climate change constitute grave security challenges that no technological fix can meet.

Finally, commonly existing psychic boundaries that separate and compartmentalize the conscious and unconscious minds, the material and spiritual worlds, the word (professed values) and the deed (actual behavior), thinking and feeling, and "masculine" and "feminine" values—all these increase the likelihood of war and freeze progress in transcending the four previously mentioned boundaries. These walls of the heart and mind reinforce authoritarian personality traits and violent power relationships, produce exaggerated fears of adversaries, inhibit flexibility and tolerance, and discourage the growth of universal identities on which future security will depend.[8]

People who do not see the gaps in their picture of reality clouded by these five boundaries more readily employ violence against other humans and disregard nature.[9] As the preceding chapters demonstrate, tendencies toward exclusivistic identification with one's nation, economic class, generation, and a single species, plus psychic habits characterized by separation and compartmentalization, distort people's perceptions and consequently limit their vision and empathy (White); unnecessarily constrict their political and technological decisions (Davis, Lynch); stifle a more efficient, less exploitative, and potentially stabilizing global political economy (Dumas); stifle the growth of robust legal norms, procedures, and institutions (Weston); unravel the weaving of more representative institutions and a more governable social fabric worldwide (Russett); and even warp people's most basic moral calculations (Coffin). On the other hand, by intentionally striving to transcend these boundaries of thought and behavior, people can envisage more powerful and effective security policies for the future. This envisaging of a more secure and compassionate world calls forth the deepest meaning in the foregoing chapters and underpins all serious discussions about alternative security.

IDENTIFYING GUIDELINES FOR ALTERNATIVE SECURITY POLICIES

To grapple deftly with these five boundary problems, a principled approach to security policy is needed. Explicitly announcing central principles or generalized intentions helps give overall direction and constancy to any nation's security policy. Once principles are publicly embraced, officials are less tempted to dismiss them for short-term opportunistic reasons. During electoral campaigns, discussion of positive principles protects candidates and the public from unscrupulous appeals to the most chauvinistic, xenophobic tendencies of the electorate. Endorsement of a principled alternative security policy by the public and by legislators also can shape legislation and constrain executive officials from deviating easily or covertly from what a more principled policy would consider prudent in the long run.

A principled policy, especially if widely known and increasingly honored throughout the world, also can serve as a corrective to the problem of "global partisanship." In a common security policy, where the entire species is considered the relevant constituency for decision-making, any national perspective, however bipartisan from the standpoint of domestic politics, is partisan from a global perspective. If electorates and officials become more aware of this problem because the same principles for security policy become touchstones for many countries, disputes are less likely to lead to violence. The use of principles as the basis for negotiations also aids enormously in settling potentially violent disputes.[10]

By articulating and honoring principled goals for an explicit code of conduct, one nation can encourage and firmly press other governments to follow suit. International judgments by many nations, for example, articulating the same principles of fair play can more effectively moderate nationally partisan policies emanating from, say, Moscow, Tokyo, or Tripoli than can Washington acting, as inescapably it must, as a mere party—not a neutral judge—in a dispute with any other country. Also, where a principled approach is rejected by one's government, individuals and non-governmental organizations may nonetheless still be able to guide their own behavior—or at least their discussions with fellow citizens—by a new international morality that rises above the more traditional policies their national governments try to implement. Such a moral code reduces stress and expands strength[11] in a social fabric that can increase the governability of the world.

Five basic principles or goals, informed by the five boundaries that impede movement toward common, comprehensive security, underpin a new international morality: demilitarization, reciprocity, equity, environmental sustainability, and accountability. Each is examined here in light of the preceding chapters; but only the first, because of its special relevance, is analyzed in detail.

Demilitarizing World Society

The purpose of global demilitarization is to reduce the role of military power in international relations generally, in contrast to arms control which aims to limit weapons of one sort or another. Demilitarization may well include gradual reduction of arms, but it goes much deeper than arms control, which, as it has been practiced since the 1950s, has meant merely the management of the arms buildup. Washington officials have departed little from the arms control approach despite the promise of the INF treaty and the enormous changes in the Soviet Union and Eastern Europe that virtually have taken away "the enemy" for the United States.[12] In contrast, a demilitarization policy would implement concrete proposals leading to the eventual reduction of armed forces everywhere to the lowest level possible while still maintaining domestic peace and monitoring borders for purposes of regulating trade and immigration. It would be accompanied by the growth of world monitoring and security institutions, such as outlined in the preceding chapters, particularly in the contribution by Burns Weston, to replace over time the need of governments to rely on their armed forces for security.

Given the comprehensive nature of the security concept proposed here, demilitarization understandably extends beyond reductions of armed forces to policies designed to demilitarize the national and global economy, public education (to reduce denial, projection, and Good Guys/Bad Guys thinking in its content), and the social, psychological, and religious habits that support the aging war system of international relations. Separatist, national military cultures would be gradually displaced by a culture of legal obligation and peaceful dispute settlement among diverse nationalities, able to flourish safely because they are protected by international human rights guarantees and no longer wedded to military power.

Assessing the Utility of War

Although most of the authors of the preceding chapters are sympathetic to a broad-gauged demilitarization policy, they may at points acquiesce too easily, albeit reluctantly, in a continued reliance on national military power for security. In my judgment, their analyses could profitably go farther toward considering systemic alternatives to war itself. An unflinching look at reality does not warrant retention of the war system, even with many reforms, beyond a transition period measured in decades, not centuries. Preparation for the transition, already pressed upon us by technology, must occur now if change is to be informed by reason and if it is to occur by transcending rather than deepening the boundaries that now threaten our safety.

The extent to which one can justify continued reliance on military power for security should be determined, if we are to be rational, by the utility of war. Yet a plausible case can be made (1) that war is not useful, (2) that it is not necessary, and (3) that preparations for war, which continue in many countries, undermine common, comprehensive security far more than they enhance it. If a sufficient body of evidence sustains these hypotheses, then alternative security thinking should not be satisfied with merely moving toward a minimum nuclear deterrent, nor even with the more ambitious task of abolishing nuclear weapons, but instead with abolishing war itself. Let us examine briefly each of these three propositions.

1. Is War Useful?

Even the most heavily armed countries have found their military power to be largely inapplicable for enhancing security in today's world. In so-called small wars against determined nationalist movements, both the United States and Soviet Union have suffered severe reversals. Vietnam and Afghanistan symbolize the ineffectiveness of vast military strength in a hostile environment against determined resistance. In both cases, the superpowers could hardly have achieved less through strictly non-military means than they achieved in their extremely costly applications of military might.

The prospects for benefits from larger wars are even bleaker; the likely destructiveness of major nuclear combat to one's own nation, one's allies, and the global environment cannot be justified by *any* conceivable political objectives. Both the US and Soviet presidents have said that nuclear war has no utility, can have no winner, and must never be fought. Influential religious organizations have called into question not only the use but also the possession of such weapons. Sensible responses to the threat of nuclear holocaust do not require fighting prowess; they require effective peacemaking to ensure that conflicts do not move to the brink of war.

A third category of violent combat, illustrated by regional conflicts such as the Iran-Iraq war and the East-West confrontation in Europe, includes wars that have become, quite simply, too destructive to constitute rational means for achieving desired ends. Even if a war in Europe could be limited to conventional arms, for example, these weapons are now so destructive that the conduct of the war would destroy the very populations whose protection is sought.[13]

The declining utility of conventional war has become apparent to West Europeans, who, through the European Economic Community, have ended the age-old war problem between Germans and French[14] and now are trying to fashion new and somewhat analogous approaches to enhancing East-West security. The declining utility of war is clear also to Soviet officials. And in the United States, time-honored realists who are independent thinkers have been so moved by an honest appraisal of the probable devastation of major war that they flatly say that not only nuclear combat but also conventional war among industrialized powers must be forever ruled out.[15] Medium-sized wars also always run the risk of drawing in other countries, as occurred in the US flagging of Kuwaiti tankers in the Persian/Arabian Gulf, with the danger of escalating conflict. But most importantly, these conflicts are really not amenable to solutions brought by violence.

Are there exceptions to the rule that war has lost its utility? The US invasions of Grenada and Panama and the bombing of Libya may seem like exceptions to the rule, although on closer examination these actions may not in fact have served US security interests in the long run. In any case, alternative means were available for dealing with most if not all of the issues raised in each instance.

In the aerial attack on Libya, Washington associated anti-terrorist efforts with US geopolitical and military interests and with anti-Libyan and anti-Arab sentiment. As a result, the policies drew criticism even from European allies and stood in the way of gaining the widespread support that a less belligerent anti-terrorist policy could probably have achieved and that eventually will be required for anti-terrorist diplomacy to be effective.

In Grenada, US security was not in jeopardy. If any direct threats to the security of neighboring nations had arisen, they could have been met by multilateral peacekeeping operations. The invasion undermined an important interest the US has in maintaining a peaceful world order by upholding the principle that no government is allowed to intervene with military force to change another country's government. One such intervention can prepare the way for another.

Essentially the same points can be made about the US invasion of Panama to oust Manuel Antonio Noriega. However deplorable General Noriega's policies, there is little doubt that Panama never constituted a threat to the United States. In the absence of a direct attack or a serious threat of direct attack on the United States or its significant interests, Washington could not legitimately claim to be acting in self-defense—the only justification the UN Charter explicitly allows for the use of force. As in Libya and Grenada, the unilateral use of force probably undermined the US long-term interest in a peaceful, law-abiding world that is congenial to US security interests in the long run.

In sum, collective violence has lost its utility in small, medium-sized, and large-scale war.

2. Is War Necessary?

Of course, critics argue that, although its utility may be low, war still may be necessary in some unusual circumstances. To determine whether this argument makes sense, consider the day-to-day security needs of the United States, recalling at the outset that it is not threatened by attack from anyone. Nonetheless, the United States and the world community have these security needs:

- to patrol and pacify tense borders in order to prevent brushfire incidents from erupting into war and to discourage the clandestine movement of arms and armed forces across international boundaries of countries where such problems have in fact arisen, *e.g.*, Nicaragua, El Salvador, Honduras, Lebanon, Israel, Jordan, Iran, Iraq, Afghanistan, Angola, Namibia, Chad, Cambodia, and elsewhere;
- to discourage intervention by external powers in local and regional conflicts, such as the US and Soviet roles in Vietnam and Afghanistan, the Cuban presence in Angola, and the Vietnamese intervention in Cambodia;
- to prevent aggression by medium or small powers against one another, as Iraq against Iran, or Libya against Chad, or Middle Eastern countries against one another;
- to prevent ethnic or religious antagonisms from becoming violent, as between Greek and Turkish Cypriots;
- to implement peaceful settlements, such as the Israeli withdrawal from the Sinai and from part of Lebanon;
- to monitor arms agreements impartially and to publicize any violations in order to discourage them and to specify the nature of the violations so as to avert overreaction by a rival, as occurred in disputes over compliance with the SALT II treaty;

- to discourage the proliferation of weapons of mass destruction in the arsenals of countries already possessing them and of countries interested in obtaining them;
- to prevent space from becoming another source of military threat; and
- to provide impartial auspices to mediate or adjudicate disputes before they erupt into violence.

Not one of these needs can be met by more US military might or by unilateral US military pressure. If the Cubans have begun leaving Angola, it is because of multilateral political settlements on Namibian independence, UN monitoring of the agreement, and an end to other nations' military intervention in Angola. If clandestine gunrunning does not occur in Central America it will be because of impartial UN and the Organization of American States (OAS) observers at the borders and negotiated political settlements, not because of well-financed "Contras." If the Soviet military buildup is reversed, it will be because of non-military priorities in Moscow, detente, arms control, and international inspection, not because of US military threats.[16] If war between countries like Iran and Iraq is prevented, it will be because the arms trade has not inflamed regional antagonisms and because the economic incentives for peace are greater than those for war; and if such wars do occur, external powers will need to avoid their escalation through conflict containment, dampening arms sales, and possibly the international protection of neutral shipping, not through introducing more armed forces in the region. If Pakistan, Brazil, and other countries do not manufacture nuclear weapons, it will be because the nuclear arms buildup of the superpowers has stopped and truly effective international rules against proliferation are enforced in every single country on earth.

By their nature, in sum, the day-to-day security problems most likely to prove nettlesome cannot be met adequately by national governments acting separately or unilaterally, no matter how strong they may be militarily. Modern security problems are made worse by unilateral uses of force. To manage them efficiently requires multilateral diplomacy and UN institutions that discourage external involvement by individual nations. Of course, the availability of multilateral peacekeeping and its imaginative extension in the future are additional reasons that war is not or should not be necessary. Peacekeeping under UN auspices, although certainly no panacea, can be more successful in serving most security needs than can uses of force by one nation or bloc of nations whose claims to legitimacy invariably are open to challenge.

Short-term security needs, then, can be well met by closing the door on national uses of military force and opening the door more widely for UN peacekeeping. True, there are some *long-term* security needs that UN peacekeeping, as presently practiced, cannot meet, the most obvious being the elimination of the danger of nuclear war. Others include the proliferation of non-nuclear weapons of mass destruction or the delivery of suitcase bombs. Yet national military forces cannot provide effective defenses against these threats either. Such security needs can be addressed in the long run only by policies that encourage global society to transcend the five boundaries noted above and to develop international norms for global demilitarization that are enforceable. This is what alternative security policies are designed to accomplish.

3. Planetary Militarization and Its Consequences

As the preceding chapters make clear, preparations for war continue despite the declining utility of war and reduced military spending and deployments by the Soviet Union. In Washington, rather than developing a program for global demilitarization, for the prohibition of intervention in regional conflicts, and for the curtailment of covert operations, officials are implementing policies to develop more sophisticated nuclear weapons and new instruments for military intervention. These priorities are reflected in a report from a leading group of US national security managers, chaired by Fred C. Iklé, which concludes that the proper course for US policy is "discriminate deterrence," thus justifying the new weapons that the Pentagon wanted even before Gorbachev's more progressive initiatives (although now to be procured at a somewhat slower pace).[17] The report encourages the United States to continue toward four goals: (1) developing new nuclear weapons and delivery systems, with greater speed and stealth, to give commanders more precision and ability to discriminate in the combat use of nuclear weapons; (2) increasing US ability to intervene quickly and decisively in "low-intensity conflicts" in the Third World; (3) preparing an expanded role for covert operations; and (4) proceeding with the development of new weapons for warfare in space.

In accordance with this strategic direction, the Bush Administration is pushing forward with the development of space weapons; and, although Congress has authorized a smaller program than the Republican administration wants, Congress has refused to press for guarantees that space will be kept weapons free. Moreover, as a result of US insistence on keeping the nuclear door open, India, Israel, Pakistan, and South Africa are rapidly enlarging their stockpiles of nuclear weapons material, with the first three trying to move from fission-based atomic weapons to far more powerful fusion-based devices.[18] Argentina and Brazil can now cross the nuclear weapons threshold at will. Iran and Iraq, which are rearming to the teeth, already possess chemical weapons and missile delivery systems, and presently are eyeing nuclear weapons. Iraq, at least, seems to have taken the first steps in a program to build them.

In addition, the proliferation of non-nuclear weapons of mass destruction, especially chemical and biological weapons, proceeds at a quickened pace. The production of chemical weapons can occur throughout the world in laboratories originally constructed for the manufacture of herbicides and pesticides. Again, conventional military techniques are incapable of countering such threats. Only global constraints, multilaterally verified, can succeed in halting the further development of weapons of mass destruction.

The technology for producing aircraft and missiles has also been spreading ominously. The world's six most advanced newcomers to nuclear weapons technology—Israel, India, Pakistan, South Africa, Brazil, and Argentina—already possess advanced aircraft and are in the process of developing missile delivery capabilities if they do not already have them. China has sold a number of its missiles, with a range of 2,200 kilometers, to Saudi Arabia. And more than twenty Third World countries currently possess ballistic missiles or are making serious efforts to develop them.[19]

The growth and further spread of military technologies profoundly endanger global security.[20] Yet the prospects for stopping, for example, the further spread of advanced technology for rapid delivery of conventional, chemical, biological, and nuclear warheads are exceedingly dim in the absence of serious steps to achieve a universal ban on ballistic missile testing. Yet such a ban cannot occur, nor can the spread of other advanced military technology be halted, as long as the major powers fail to demilitarize their own security policies.

Thus, preparations for war continue, undermining everyone's security in the long run.[21] Given the imperfect functioning of the existing balance of military power and the proliferation of military technologies worldwide, we reasonably can expect that war some day will come again—unless, that is, we transform the structure of international relations to provide sufficient governance at the global level to strengthen multilateral peacemaking and peacekeeping while negotiating and enforcing the demilitarization of national war-making potentials.

Abolishing War

The failure to be sufficiently impressed with the inutility of war and of preparations for war has led many analysts to focus too narrowly on the dangers of nuclear deterrence. The focus is understandable because nuclear weapons are, of course, the most fearsome and least justifiable weapons in military arsenals. Yet, at its best, alternative security aims not at living without nuclear weapons; it aims at living without war.[22]

Unlike many exponents of alternative security who seem willing to live with nuclear deterrence at minimal levels,[23] I think it is unrealistic to believe that the threat of nuclear weapons can be reduced to a tolerable level by establishing a minimum nuclear deterrent, although admittedly it would be a useful step toward the abolition of nuclear arsenals. If the industrialized countries with nuclear weapons claim, as they do, a right to keep their weapons indefinitely, they cannot expect other less secure and more needy countries to give them up entirely and forever. This expectation, I speculate, simply will not be fulfilled. Either all countries must give up their nuclear weapons or all countries must accept that any country that wants them eventually will get them.

No amount of glossing over this stark reality should be allowed to divert those working seriously for an alternative security system. Advocates of demilitarization must examine carefully the widely held yet questionable belief that the world would be less safe with the abolition of nuclear weapons than with the maintenance of nuclear deterrence. This belief is based on one or both of two assumptions: that nuclear weapons have kept the peace for four decades and that someone could cheat on a total ban. Whereas hiding a few nuclear bombs is militarily insignificant in a world of thousands of weapons, hidden weapons could be exceedingly dangerous, the argument goes, where all countries except the cheater have dismantled their nuclear arms.

Although space does not allow a full development here, a plausible case can be made that these assumptions are erroneous and that efforts to abolish all nuclear weapons are desirable. The case is based on the following four central arguments.

First, as John Vasquez has demonstrated, there is no logical or empirical foundation for the belief that nuclear deterrence has prevented nuclear war.[24] His analysis, as well as the recent work of other scholars,[25] shows that the central element of US security policy over several decades may have had little to do with eliminating what was perceived to have been the main threat to US security.

Ralph White emphasizes that *military* deterrence, a much broader concept than *nuclear* deterrence, has prevented war in some cases.[26] His is not, of course, an argument for the indefinite continuation of nuclear deterrence. Yet even the broader concept of military deterrence is not as compelling as often is presumed. Ralph White, for example, cites evidence that military deterrence prevented Hitler's Germany from making war between 1933 and 1938. Yet he may dismiss too easily evidence that military deterrence also brought a form of confrontational, punitive diplomacy and unfriendly international economic policies toward Germany that created the political conditions that brought Hitler to power in the first place and that encouraged him and many Germans to make war in 1939.[27] In other words, military deterrence did not succeed. In contrast, if the international community would have de-emphasized military deterrence in favor of more conciliatory policies, such as proposed here, from 1919 on, it might have prevented the conditions that facilitated Hitler's rise to power or averted World War II altogether.[28]

Ralph White believes that the case for deterrence against Hitlers of the future is quite strong.[29] Yet if a future Hitler possessed nuclear weapons, would the world be safe, regardless of how many other nations had them also? What would be an appropriate US response to a future Hitler who might, during a showdown of threats, proclaim willingness to engage in combat to the brink of "nuclear winter," perhaps because of desperate conditions at home, unless certain "legitimate" wishes were met? If a future Hitler had nuclear arms, the threat of others to use their weapons against such a person might not be credible. Alternatively, their threat might be credible, but the future Hitler might not believe it to be. Either way, nuclear war would result. If we generalize Hitler to be the stereotypic "crazy leader" or "mad Caesar," nuclear deterrence offers no hope at all, because deterrence assumes a rational, not a crazy opponent. If an opposing leader does not calculate rationally what other governments will do, or if that irrational leader does not agree with what others consider to be unreasonable levels of damage that ought to deter misbehavior, then deterrence fails.

Therefore, the international community must prepare defenses *other than military deterrence* against future Hitlers. It must ensure that future Hitlers are unlikely to come to power, that they are unable to obtain nuclear weapons if they do come to power, and that they are unlikely to garner sufficient support from their own government and population to launch a suicidal war, because the firm yet non-threatening intentions of the rest of the world would be unequivocal. These goals can be achieved only by establishing a total ban on the possession of nuclear weapons in particular and by implementing alternative security policies in general.

Second, to the uncertain extent that military deterrence does reduce the likelihood of war, such deterrence can be carried out, at least against a government secretly trying to violate a nuclear weapons ban during a limited transition period, without resorting to nuclear weapons. Non-nuclear weapons are sufficiently destructive and precise to destroy any society on earth, and they can do so without nuclear fallout.[30]

Third, cheating is not as large a problem as often claimed because intrusive inspection is becoming politically more acceptable and because intense developmental efforts can improve technological abilities enough to provide high confidence of monitoring all societies. In addition, modern communication, combined with individual citizens' willingness to report possible violations to a designated international monitoring authority, can produce high confidence of verifying a universal ban on nuclear weapons.

Fourth, the overall risks of nuclear proliferation, which attend any global posture less stringent than total abolition, exceed the low risk that any militarily useful secreting of nuclear weapons could occur without detection in a gradually demilitarizing world.

There is yet another reason for focusing more heavily on the war problem than the nuclear deterrence problem. Even if one seeks the total abolition of nuclear arms, it seems unrealistic to expect that the present members of the nuclear club will give up their nuclear weapons as long as conventional war against more populous adversaries remains a serious possibility. Even if additional countries do not rush to build nuclear bombs, the spread of chemical weapons, sometimes considered the "poor man's bomb," poses another frightening prospect that makes programs of partial arms control ineffective. Moreover, even if, for argument's sake, it were possible to dismantle all nuclear arms, nuclear knowledge still would pose deep-seated fears that some government might quickly rebuild such weapons during moments of tension—assuming, that is, that an effective global enforcement system had not already been established to prevent this possibility.

Thus, the nuclear sword cannot be lifted from our heads unless two conditions are met: (1) all countries must become permanent nuclear-weapon-free zones; and (2) all countries must drastically reduce their offensive, conventional military capabilities so as to preclude the possibility of large-scale offensive military operations. In our attempts to implement these conditions, the important focus must be not on any particular weapon or even class of weapons, but instead on the gradual, unrelenting, reliable reduction of the role of military force in international relations generally, until war itself has been abolished as an acceptable institution.

Restructuring Sovereignty

To focus on all forms of collective violence, rather than on nuclear deterrence alone, also highlights two other important points. First, the nature of the war-making function of sovereignty is changing. Indeed, it has changed so much that to retain any longer in national hands the ability to make war independently of any higher authority is to give up the overriding purpose of sovereignty itself, namely the ability to achieve the security of the nation. Second, the location of the residually legitimate function of the war-making dimension of sovereignty—police enforcement—must move to the global level, the only place where it can be reasonably carried out.

Apologists for traditional sovereignty in the balance of military power need not fear that they would be losing something that, at reasonable cost, they could retain anyway. Technological advances will continue to transform the nature of sovereignty, making separate, independent decisions less possible. Efforts to retain a war-making function of national sovereignty much longer actually will risk its extinction for many, if not all, nations—through actual war or intimidation—rather than bring its transformation into police enforcement for the good of the global community.

The Moral Question

Above all, the moral issue must be faced more directly than any government and most people have done thus far. If nuclear weapons ever are used widely in combat they can only be characterized as genocidal and ecocidal weapons, indiscriminately inflicting death on many innocent people and the planet's life-support systems. In looking back now on the genocidal instruments of the Nazis, people almost universally conclude that individual Germans should simply have said "no" to paying for gas chambers, to constructing ovens, to operating these instruments of death, and to tolerating their very existence.[31] Genocidal weapons, whether gas or nuclear, should be recognized as evil incarnate. One country's genocidal weapons do not justify another's. In any society where human rights are honored, no citizen should be required (or agree?) to pay for, build, transport, or tolerate the existence of instruments of genocide.

Non-Military Deterrence

While one side of a demilitarization program constrains the role of weapons, the other side increases the strength of non-military influences, as illustrated in the chapter by Burns Weston in this volume.[32] Non-military forms of dispute settlement, such as international regimes, legal tribunals, and boards of mediation, gain strength with frequent use. Diplomacy can be seen as an educational process in which every policy provides teaching and learning possibilities for the international community. The establishment of a permanent, individually recruited police force, for example, has an educational purpose that at this stage in history is as important as the physical security it can provide. Such a force could help people to understand that the global enforcement of community norms on behalf of the community is indeed possible. A permanent, individually recruited UN police force could establish a sense of global identity and impartiality that *ad hoc* forces made up of donated contingents from national armed forces cannot do as well. The agenda for building world security institutions includes also a global monitoring and research system, a non-interventionist security regime for smaller countries that could be protected by the UN, an environmental "security council" without veto, and other specific suggestions made in the preceding chapters in this volume.[33]

The political flux in Europe presents historic opportunities to demilitarize and to create new common security structures in Europe. Yet, as of this writing, the Bush Administration plans to retain the East-West alliance structure and to bolster NATO by giving it some non-military activities. For the administration to ask that Germany remain within NATO, even if West and East Germany were to be unified, is a throwback to old, divisive habits of trying to obtain a military advantage through a peaceful gesture. Such policies are less useful than would be Washington's support for a plan to create an all-European security system in which the association between two parts of a permanently denuclearized and substantially demilitarized Germany would not threaten any country.

Demilitarization: In Whose Interests?

Exponents of alternative security need to pay close attention to the structure of interests and forms of learning that advance demilitarization. In this regard, they differ from advocates of traditional security policies who place almost all emphasis on military power. Also, they differ from the advocates of world federation in the 1950s who failed to develop an effective political strategy. A major problem in building consensus for change—in North, South, East, and West—is the failure to focus sufficiently on the need to develop concrete diplomatic and educational programs aimed not so much at policy reform as at the more fundamental task of transforming the international system itself. Current change efforts tend to be piecemeal, haphazard, and half-hearted. If one or two countries negotiate arms reductions, or if several countries seek to curtail the arms trade, or if one region seeks to establish a nuclear-free zone, but meanwhile the world continues to shy away from real reductions in the role of military power in the international system, even positive, incremental measures are bound, eventually, to fail.

Officials and publics need to recognize that arms are less the problem than the willingness of governments to use them and the failure of societies to build non-military means for security enhancement. Citizens and governments will not give political support to reducing their own arms very far if they harbor fears that other nations, even in the distant future, may threaten them. The US public's current receptiveness to arms reductions, for example, does not mean that it trusts the Soviet Union. On the contrary, at the same time that more than two-thirds of the voting public would like to move toward elimination of nuclear weapons, 68 percent continue to believe that "we cannot trust what Soviet leaders say, so we should proceed slowly and with caution. . . ."[34] Similarly, the public will not support substantial arms reductions if they believe that other nations, even though small and far away, may someday bring war closer. Although most US citizens, for example, no longer expect a direct nuclear attack on the United States by the USSR, 60 percent believe that smaller countries such as Pakistan, Israel, or South Africa "will eventually use nuclear weapons."[35]

The solution to this problem is to develop and support a comprehensive plan that aims not merely at reducing arms but at reducing the role of military power in world affairs. To reduce arms and simultaneously take account of widespread fears of one's adversary, as is politically essential for every leadership to do, means that world security institutions, especially as outlined by Burns Weston,[36] are necessary to help transcend the adversarial nature of today's security system and to provide additional reassurances that security can be enhanced by dependable global mechanisms. A successful plan must emphasize the growth of world security institutions as much as arms reductions so that alternative means for security enhancement are being erected at the same time that arms are reduced.

One of the most promising developments, of course, has been the series of remarkably progressive initiatives taken by Mikhail Gorbachev and the Soviet government. Almost without recognition by US officials, Moscow is asking the United States and the other great powers to strengthen international institutions and to revamp the customary code of international conduct so as to permit the mitigation of the pressing global problems that no national government can handle separately. Georgi Shakhnazarov, one of Gorbachev's closest special assistants, has given an extensive pragmatic rationale for Soviet determination to restrain traditional national interests that heretofore have disregarded the human interest.[37] Writing in *Pravda* and elsewhere, he argues that Soviet policy must be guided by the need to raise the "governability of the world" to levels where global governance can manage global problems. Gorbachev's policies, he explains, ask "every member of the world community to place universal interests above class, nation, group, ideological, or other interests." In a statement that is as remarkable as it is overlooked by the West, Shakhnazarov observes that even some of the main arguments against world government, which made sense for several decades after World War II, are "no longer there." In sum, Gorbachev's initiatives seek to bridle today's balance of military power with a growing web of international laws and organizations, until they function instead as a legally constituted balance of *political* power. In such a system, military power would be gradually relegated to a smaller and smaller role until war itself would be ruled out as a technique of conflict resolution.

Gorbachev's initiatives have put many internationalist items on the world's agenda that have not been there in years, if ever before. Creative space clearly exists in Moscow, not merely for reducing arms, but for the far more important goal of reducing the role of military power in international relations generally.[38] Taken together, Soviet proposals and deeds offer more opportunity to institutionalize law and order in world affairs than has existed ever before in modern history, even more than what accompanied the close of World War II when Washington, Moscow, and the other Allied Powers created the United Nations. To this can be added the strong support for arms reductions, if not demilitarization, given by the non-aligned countries for many years. Repeatedly these countries have pressed for major steps in disarmament, including a comprehensive ban on nuclear testing. Moreover, Bruce Russett suggests that great promise may lie in the growing number of non-aligned societies in which democracy has taken root; fewer military governments means that more States will relate to each other without war.[39]

Within the United States, which has tended to resist many arms control initiatives, especially since the beginning of the Reagan Administration, Thomas Lynch emphasizes the extent to which professional military people can be brought into the alternative security equation, by giving them important functions in defining new security structures.[40] That is wise counsel as long as the goal remains clear: to dismantle—not to remodel—the existing war system in international relations. He also cautions that "no realistic alternative to nuclear deterrence can or should evolve without the active engagement of the world's militaries in its definition and execution."[41] It is of course important to include the military in changes that affect them directly. Yet is it accurate to suggest that no realistic alternative to deterrence can evolve without the active involvement of the military? Are not alternatives evolving already, with every person, church, municipality, and country that rejects nuclear weapons helping to bring these alternatives closer to the foreground? If sufficiently widespread, might not these efforts convince others to accept a nuclear-free outcome, even though their fears prevent them at first from embracing it on their own? Surely supporters of alternative security need not wait for the architects of nuclear deterrence to dismantle it any more than slaves and abolitionists needed to wait for plantation owners to dismantle slavery. To be sure, people should be civil toward their political opponents, but they need not delay pressing for major change until the last holdout participates in planning the change. Do "sovereign military forces"[42] have as much right to threaten societies with instruments of genocide as a sovereign people have the right to insist that genocidal weapons be abolished?

Thomas Lynch seems to argue that, even if the military arguments against alternative security have little logical basis to them, exponents of alternative security should scale down their proposals by not calling for actions that would drastically cut military budgets or personnel. He feels that such cuts will arouse the powerful military bureaucracy to oppose them. True, it may be easier to demilitarize the military than to dismantle it. But perhaps Thomas Lynch's argument concedes too much. If military officials cannot justify on security grounds everything that they spend and deploy—as, for example, they could not do in deploying multiple, independently-targetable re-entry vehicles (MIRVs)—then no one serves the country's security by moderating policies just because the military bureaucracy might oppose them. As occurred in the MIRV case after Moscow deployed multiple systems also, the result would be lost security.

Indeed, Paul Kennedy has demonstrated the folly of such a strategy even on strictly economic grounds, showing that mature great powers and their military bureaucracies usually make the *wrong* decisions for their own good. They overextend themselves economically to procure military power until eventually they undermine their own strength.[43] The policies of the White House and Congress over the past decade, running up the largest deficits in history primarily to buy unusable weapons, seem to confirm that, unless it alters its course, the United States will not save itself from this historical pattern of unnecessary demise.

The burden of proof, then, must be on those who propose more military equipment to demonstrate that their proposals do not undermine security. Militarization programs must be compared with demilitarization programs to determine their relative contributions to the full range of security problems, including the threat of economic losses, nuclear proliferation, global warming, and the prospect of "nuclear winter" if nuclear combat should occur. After lengthy study, John Tirman has reported that "breaking the grip of the military estate on America's policy-making, weapons production, and value system would go further than any other measure to reorient national security to more appropriate ends."[44]

This task need not wait for other governments or even for Washington to act. Individuals and non-governmental organizations can work at it through local schools, churches, and the media.

Indeed, alternative security thinking underscores that the control of the military by persons who do not operate from a military mentality is different from and more important than civilian control of the military. Years ago civilian control seemed a solution to averting the dangers of militarism. But today many of the civilians most commonly selected to lead the national security establishment have adopted military values. In fact, they often urge military interventions where professional soldiers dutifully agree to go but know in advance that a quagmire awaits them. Permanently high levels of military preparedness mean that militaristic minds, whether civilian or not, usually control the armed forces. They are advocates of more military power rather than of a decreased role for military power. Thomas Lynch and others give insufficient attention to this problem. As Tirman concludes, "if the grip of military values on society is not loosened, the possibilities for innovative leadership . . . essential to common security are very dim."[45]

Of course, policies that reduce the military budget and bureaucracy are bound to be politically troublesome, as Thomas Lynch correctly reminds us.[46] This is why alternative security policies include legislation to manage economic conversion. Lloyd Dumas has shown how imaginative arrangements can adjust the military's economic and bureaucratic interests without leaving people unemployed even after military budgets are sharply reduced.[47] Such proposals recognize the goodwill that exists within parts of military organizations and attempt to build political support within the military for what needs to be done. This seems more useful than moderating requests for demilitarization because vested economic and military interests may oppose them.

Military support may grow also from a recognition that there are many important future tasks that military personnel can fulfill better than others. Defense ministries could provide extensive monitoring and inspection personnel to improve the ability to verify norms restricting the manufacture and possession of arms. In addition, expanded research and development programs to construct better verification equipment are needed. Security enhancement activities would also include policing environmental standards to avert accidents at sea or in space and taking measures to clean up or otherwise recover in the wake of environmental or natural disasters. Other tasks might include volunteering for a permanent, individually recruited UN police force once established, just as peacekeeping has become a natural and growing part of the role of the Nordic armed services, for example, where military professionals have come to accept UN peacekeeping as a significant part of military life.[48] National military personnel would, of course, continue to police borders for the purpose of regulating commerce and immigration even in a demilitarizing world.

To military people genuinely committed to the enhancement of security, measures such as these would be attractive as long as it is clear that, by curtailing traditional military roles for one's own country, similar limitations would eliminate the threats that any adversary might pose to one's own society. Indeed, the military's highest duty is to do the utmost to reduce threats to the security of the homeland. Helping to implement a prudent program for global demilitarization could be such a high calling.

Honoring Reciprocal Rights and Duties

We turn now to a briefer discussion of the remaining four principles of alternative security policy. The principle of reciprocity, the first of these, is perhaps the most fundamental and already widely endorsed principle on which an alternative security policy can conveniently rest. Although reciprocity is frequently violated in practice, its universal endorsement by governments, regardless of ideology, nationality, or religion, provides a basis for attempting to hold governments accountable to a fundamental ordering principle. If rigorously applied, no government could with impunity insist on a right for itself that it would not willingly grant to others, or claim a duty for others that it would not accept for itself.

If seriously implemented, the principle of reciprocity alone could help to eliminate most wars. For example, if the United States denies that the Soviet Union or Iran, Libya, Syria, Vietnam, and other governments have a right to support armed insurrection or to finance the clandestine movement of military forces across borders, then, following the reciprocity principle, the United States could not claim that right for itself, as it has done in attempting to overthrow the Nicaraguan government. If the United States does not want to extend to other governments the right militarily to invade their neighbors to install new governments, then Washington must not claim that right for itself (even for the purpose of ridding Panama of Manuel Noriega). Alternatively, if the United States claims a right, as it has done under the Carter Doctrine, to use force unilaterally to maintain access to the oil fields of the Middle East, then may poor countries claim a right to use force to gain access to the corn fields of the Middle West or to encourage a migratory "invasion" of starving people to obtain food in a country that, by most measures of international justice, consumes a disproportionate share of the world's resources?

To install reciprocity as an operational policy principle can also drain heated ideological hostilities or religious fanaticism from many conflicts, because disputing officials can then focus on principles of good conduct rather than on the ideological goals of the other. Ideologically diverse governments can live peacefully with one another as long as they understand and explicitly design policies to respect this fundamental principle. Reciprocity and "realistic empathy" reinforce each other and provide a check on people's unconscious rejection of evidence that conflicts with "unconsciously cherished images."[49]

In addition, reciprocity provides an essential antidote to one of the most serious dangers of a principled foreign policy: the tendency to assume a moralistic attitude, based on the assumption that one's own policies are more virtuous than the policies of others. As well-schooled realists know, governments characterized by "legalism-moralism" often insist on their own way, propagandize for the "correct" point of view, and threaten other nations who obstruct their policies. Because evaluating respect for reciprocity can be controversial, as illustrated in US and Soviet claims alleging violations by the other of the ABM treaty, compliance with reciprocity can be enhanced if its meaning is authoritatively interpreted by third parties whenever possible. Moreover, the implementation of complementary policy principles, such as democratization and demilitarization, discourages the possibility that a policy inspired by a moral vision will take on a self-righteous or aggressive tone. The recommendations in this volume by Ralph White for empathy in policy-making, by Burns Weston for wider application and impartial implementation of legal norms and processes, by Lloyd Dumas for economic fair play throughout the global arena, by Bruce Russett for recognizing the peaceful potential in nurturing democracy globally, and by William Sloane Coffin for honoring the contribution of moral principle to prudential politics — all undergird the principle of reciprocity which in turn constitutes the bedrock of alternative security. Alternative security means nothing unless it includes an active concern for the security of one's adversary, because there is no path to one's own security other than through the door of reciprocal expectations that adversaries will reduce their threats in return for one's own threat reduction.

Achieving a Sustainable World Society

Without deeper respect for nature, a life of dignity for the human species cannot continue on planet Earth. Environmental issues pose planetary dangers of such a magnitude in scope and severity that they constitute the most serious long-range security problem in the world today.[50]   As the World Commission on Environment and Development (commonly called "the Brundtland Commission") concluded, life support systems for the entire human species face profound and uncertain threats from pollution, resource depletion, population pressure, and species extinction.[51]   Although these serious security problems can be addressed in part within local and national contexts, none can be treated effectively through the traditional security instruments of military strength or even through the traditional diplomatic instruments of bilateral diplomacy. People now need decision-making authority that is globally binding. Only the transfer of massive financial resources and brainpower from military to environmental purposes and only truly cooperative, binding, multilateral legislative efforts to protect the ozone layer and halt climatic change induced by global warming will enable the species to survive and enhance the dignity of life for future generations.

Although the problems seem at first to be overwhelming, the Worldwatch Institute has estimated that $77 billion a year over a single decade could reverse adverse environmental trends in four key areas: protecting topsoil from further erosion, reforesting the earth, increasing energy efficiency, and developing renewable sources of energy. This cost amounts to only 8 percent of world military spending each year, with the benefit of laying a foundation for subsequent steps to end hunger and avert global warming. Alternative security policies could redirect large portions of global expenditures on research and development for new military technologies, estimated at $100 billion in 1986 alone, into developing new energy technologies, increasing agricultural productivity, enhancing pollution control, and improving human health.[52] All the world's governments combined presently devote fewer funds to these activities than what they spend on developing new military technologies.[53]

Both the exponents and critics of alternative security policies can benefit from studying the exciting possibilities for managing environmental problems that have arisen from the willingness of some governments to reformulate the nature and relocate the exercise of sovereignty. Two dozen governments meeting at The Hague in March 1989 called for a strong international environmental institution within the UN system to render binding decisions even when unanimity cannot be achieved among all members. Disputes that could not be resolved through negotiation would be referred to the International Court of Justice.[54]   Cooperation in one security area, such as managing environmental problems, of course suggests models and establishes trust for carrying out alternative security policies in other areas, such as arms reduction and global police enforcement.

Achieving Equity

The guideline to advance equity throughout global society arises from a moral desire for more justice and a pragmatic need to achieve more economic rationality, international cooperation, and willingness to sacrifice for the good of all. A more equitable distribution of economic and political resources would contribute to development programs that meet the needs of all people, as well as reduce the political power of military establishments. In addition, it would help to deprive adversaries of the opportunity to maintain what Ralph White calls a "diabolical image" of their opponents.[55]

In part because the world community has not eliminated glaring inequities, the world economy now functions so inefficiently that it damages the interests of all nations, rich and poor alike. The North's industrial capacity remains underutilized while the South urgently needs goods the North can produce. This inefficiency helps to perpetuate the inequities that cause the inefficiency in the first place, creating a form of global apartheid in which North America, Europe, Japan, and a few other countries live in relative affluence while one-half of the species lives in varying degrees of poverty.[56] These inequities constitute a security threat for three reasons.

First, poverty and inequity give rise to violence and militarism. Military governments and exponents of covert activities, whether governmental intelligence agencies or underworld drug dealers, often take advantage of poor societies.   Richer, more powerful organizations manipulate the poor, inviting both the violence of authoritarian repression and revolution against it. Such conditions also encourage military interventions by others.

Second, poverty and inequity undermine the growth and stability of democracies. In societies facing desperate conditions, extremist political leadership or authoritarian military governments often come to rule. Poverty also stimulates unwanted population growth that would be reduced if more equitable, higher standards of living were attained throughout the world.

Third, poverty is a cause as well as a consequence of environmental decay, which in turn threatens national and international security. Poor societies try to cut development costs by accepting low environmental standards for polluters. They may destroy rain forests to earn cash, even though the forests produce oxygen and absorb carbon dioxide, essential functions for people everywhere. Poor people, in desperation, also may destroy plants and trees while scavenging for food and firewood. Their herds may overgraze marginal pastures and add to the spiral of deforestation, desertification, and global warming. Overfishing, overgrazing, desertification, and loss of topsoil contribute eventually to a declining standard of living for the species everywhere.

As a result, poverty and inequity should become security concerns for the rich even if they have little moral concern for justice or for the well-being of the poor. No national government, no matter how enlightened, can progress toward a less militarized, ecologically healthy, and politically cooperative world society unless today's gross economic inequities are overcome. Toward this end, governments should consider the following illustrative measures:

- the development of economic conversion plans to reduce domestic opposition to demilitarization from corporations and workers benefiting in the short run from military contracts and to redirect resources to achieve other security goals such as abolishing worldwide poverty, sharing burdens fairly to achieve environmental protection, and providing sufficient prosperity to reduce population growth;
- serious international programs to conserve energy and other resources for the purpose of combating hunger, disease, underlying poverty, and environmental threats to security;
- automatic transfer payments from rich nations to poor nations, as recommended by the Brandt Commission, for example, which urged that all States contribute to a capital fund for development, based on a sliding scale related to national income and payable through what "would amount to an element of universal taxation,"[57] in part because appropriate economic development can enhance national and international security; and

- more equitable representation of poor countries in the international monetary system and the development banks because such representation would help to establish more stable exchange rates, bring symmetry to the burden of balance-of-payments adjustments, expand international liquidity, deal with the debt that keeps many societies buried in poverty, and thereby contribute to the overall stability of world society.

In addition, a greater degree of equity within and among countries would increase everyone's stake in avoiding traumatic disruptions of world politics and encourage all States to support politically stable, representative international institutions, the fifth principle for framing alternative security policies.

Democratizing World Society

The *democratization* of world politics is no less important for alternative security advocates than the *demilitarization* of world politics. At the bare roots, conflicts grow into wars because some government mistreats or threatens to mistreat some people and other governments oppose the mistreatment, with arms when they deem it necessary. But if democratic processes can be nurtured at every level, from the family to the planet, we are likely to experience far less mistreatment and to enjoy a far more peaceful and economically equitable world,[58] as well as a world more respectful of human rights and nature. If human rights protections against mistreatment, whether it occurs as a result of domestic authoritarianism or international aggression, can be established with the help of more democratic international institutions, then no legitimate rationale for war would remain. Thus democratization and demilitarization reinforce each other.

Increasing Governmental Accountability

The most potent antidote to war is to increase governmental accountability to all people affected by government decisions and to construct international institutions to ensure that people with severe grievances have a court of appeal outside their own country. An effective alternative security program intentionally tailors all of its policies to implement the principle of accountability at all levels of social organization – international, national, and local.[59]

Our focus here is primarily on the importance of increasing governmental accountability to people who, regardless of their nationality, are affected by the political, economic, and environmental decisions that various governments make, regardless of the national capital in which a decision is made. In concrete terms, if the steel mills in Gary, Indiana, cause lake-killing acid rains in Ontario and Quebec, then the environmental decisions by Indianapolis and Washington to govern the mills and other smokestack industries of the Midwest must be accountable, in this regard, to the people of Canada. Yet the principle of accountability, so important to the first Continental Congress and to those who founded the United States, has been virtually ignored in recent years while a growing number of decisions that affect the lives of US citizens have surfaced in Japan, West Germany, the Soviet Union, and elsewhere.

Like it or not, to maintain a democratic way of life domestically, the democratic principle of accountability must be implemented internationally. As interdependence increases, the number of decisions that affect the citizens of one country, yet which occur outside that country, will increase also. For example, if the political, economic, environmental, and human rights decisions that affect, say, the US public increasingly occur outside the United States but representative global governance does not increase, then the degree of democracy for US citizens declines, even though domestic democratic forms, such as elections and a free press, continue to function. Of course, precisely the same problem exists for people in other countries, only for most of them the degree of democracy is smaller to begin with, either because their own domestic institutions are not democratic or because they lack the power and wealth that has enabled the United States to be well represented — arguably over-represented — in world diplomatic councils.

Most citizens of existing democracies are not aware of the extent to which they are letting democratic life ebb from their societies, because they assume that patriotism means concentrating sovereignty at the national level. Although they could, alternatively, support popular sovereignty in which people allocate political authority at local, national, and global levels according to its functional utility, many citizens are unnecessarily fearful, even hostile, to internationalist solutions to contemporary problems. As George Bush declared when campaigning for the US presidency: "I will not turn one ounce of our sovereignty or of our leadership over to the United Nations. We must continue to lead the free world."[60]

Such a proclamation reveals a false consciousness. To think that one protects democracy and freedom by concentrating decision-making power in separate national governments does not make sense in a world where political decisions and influences that affect one's life occur outside one's own national institutions. Analogously, a family may establish an internally democratic structure of relationships among its members, but if it cannot participate in the governance of its neighbors or if it happens to live in Nazi Germany, it does not live in democracy.

The physical security of everyone in the Northern Hemisphere depends, from time to time, on military decisions made in Moscow, Washington, or perhaps Tel Aviv or a Twenty-first Century Sarajevo. And in today's world the economic security of most societies depends on decisions made in Tokyo, Brussels, and Washington, as well as in their own national capitals. Alternative security underscores that the society relevant to anyone's life is now global as well as national and local. As a result, patriots of democracy deceive and potentially destroy themselves if they insist on being patriots also of a sovereignty fossilized exclusively in the territorial State. Alternative security seeks to remodel sovereignty rationally to overcome the inadequacies of existing world security arrangements.

Toward this end, most nations would benefit from establishing more democratic global institutions as soon as possible. The reliability of such institutions would then be firmly established by the time that governments almost certainly will need to depend on them. In delaying these developments because of reluctance to embrace multilateral diplomacy, the United States is pursuing a perilous course toward the year 2000, when the industrialized countries "will be home to only 20 percent of the world's people."[61] Given these realities, as well as the moral imperative to enhance respect for all peoples' human rights, it is essential to establish strong protection for minorities wherever majorities rule, whether nationally or in evolving global institutions. Such protection will encourage minorities to accept majoritarian procedures globally and help to dampen hostilities among ethnically diverse populations nationally. Perhaps the only way that people in fragmented multinational societies will be able to handle their nationality problems fairly and non-violently is (1) to institute international guarantees for minorities and (2) to reduce, through global democratization and demilitarization, the differences that ethnic groups seeking more autonomy might experience between, on the one hand, having their own independent nation-state and, on the other, constituting a province of a national federation or confederation that operates responsibly in a law-oriented world community. The implementation of these two conditions would also enhance international peace.

Fortunately, instruments to protect important minority rights exist already in the UN human rights covenants and other human rights treaties. If ratified, they enable citizens to bring grievances against their own governments within designated international settings for possible redress of injuries.[62] The United States should not delay the ratification of these treaties any longer. Once a party, Washington could more effectively support efforts by the UN Human Rights Commission, regional institutions, and non-governmental organizations like Amnesty International to monitor the performance of all governments. The advancement of human rights can of course stimulate additional political strength among persons who are committed to alternative security policies throughout the world but whose influence has not been previously felt.

Expediting Global Decisions

Stronger, representative global institutions are needed also because traditional diplomatic approaches take too long to produce important decisions. Prolonged negotiations, often hampered by the veto power of some States or the non-ratification of treaties by others, limit the world community's ability to anticipate crises before they become acute and to ensure a globally unified response in relation to them. Yet global environmental threats, to take only one example, increasingly need to be anticipated and managed before they occur. In the military sphere, anticipatory decisions to control space weapons, radiological weapons, and other weapons of mass destruction require building international consensus and truly regulating what all nations do militarily. Muddling through with traditional balance-of-power procedures is too crude and weak an approach to ensure reliable security policies.

Ending Covert Interventions

Alternative security thinking also aims at eliminating covert activities. Stuffing ballot boxes, manipulating elections, bribing officials, assassinating unwanted political leaders, plotting *coups d'état*, and funding "secret" wars — all such activities are unnecessary for security and unjustifiable in a principled security policy. They violate democracy and self-determination. They often contribute to violence. As for reciprocity, what country would allow these activities to be conducted in its own or its allies' territories? If some international interventionary action clearly is warranted to rectify a desperate situation, then it can be taken openly. If the need arises for international policing, this could be done through multilateral peacekeeping. Unlike covert interventionary operations, the covert gathering of intelligence might well continue. But with the creation of an effective global monitoring and research agency within the UN system, no government would need to rely as heavily on its own intelligence sources as it has in the past.

Nurturing "People Power"

The breathtaking successes of "people power," most recently in Eastern Europe, are another source of encouragement for building new coalitions of political support to implement alternative security policies. Recent progress in removing autocratic governments around the world can, if properly nurtured, enormously increase the prospects for institutionalizing more democracy at global levels of governance. World peace will benefit because authoritarian governments are less likely to serve the interests of their people, including their security in the broadest sense, than are more popular, responsive governments. In short, the spread and strengthening of "people power" is conducive to global security.

As Bruce Russett has shown,[63] democratic societies appear to nurture more inclusive forms of identity toward each other than they do toward authoritarian societies. At least between democratic societies, progress in democratization is a way of combating "pseudo-speciation" and ending people's exaggerated fears of opponents that often lead to conflict and war. Democratic procedures facilitate the growth of realistic empathy. They enhance people's sense of inner strength as individuals and make spiritual renewal a genuine possibility. The contagion of anti-autocratic movements that swept Eastern Europe during 1989-90 illustrates the power contained in people who decide to exercise responsibility for the shape of their government.

The power of civilian resistance as a means of national defense, although it is different from mass action to oust an indigenous authoritarian government, gains some credibility from the way that repressive governments were removed from Eastern Europe in 1989. In assessing the utility of "people power," Thomas Lynch claims that it could not succeed against ruthless governments.[64] Yet in both East

Germany and Romania, where the respective heads of government ordered well-trained security forces to kill as many demonstrators as necessary to "restore order," the governments were later forced to abdicate. The order never was carried out in Leipzig after lower-ranking officials, confronted at the scene with hundreds of thousands of protestors, reluctantly reversed a written order of then President Erich Honecker.[65] In Romania, the armed forces, after at first participating in the shooting of unarmed demonstrators, refused to continue the slaughter and turned against their own commander-in-chief and his personal security guards, who continued killing as many as they could.[66] Clearly, the power of unarmed mass resistance, even against entrenched, ruthless governments, has been underestimated.

In addition, many critics, Thomas Lynch included, mistakenly assume that civilian resistance should be judged on the basis of whether it alone can do the work of an entire security policy. It never should be considered in isolation from other aids to security. When combined with significant steps toward demilitarization and the growth of world security and human rights institutions, it can be a vital asset in both maintaining peace and ensuring the responsiveness of governments that may be tempted to stray from accepted behavioral norms. It is not necessary, as Thomas Lynch seems to suggest,[67] that "people power" must demonstrate that it can stop a missile attack in order for it to be worthy of consideration. After all, even the US armed forces in their entirety cannot repel a missile attack from a determined aggressor.

The limited attention given to the potential of mass protest as an instrument for the promotion and protection of important values suggests that many discussions of security focus too much on government-to-government relations. It may be important in the future to think more about pressures that people can exert directly on their own governments, whether to constrain military activities or to protect human rights. Transnational coalitions of organized citizens groups working together and in tandem with international peacekeeping and monitoring organizations, for example, could play a useful role in bridling the excesses of the existing balance of military power. In some ways, the efforts of citizens, encouraged by Mikhail Gorbachev's new thinking, have done more to set the stage for globally significant, positive changes in democratization, demilitarization, and reciprocity than have several decades of military alliances confronting one another.

Forming transnational alliances with non-violent demonstrators and building effective international support for the rights of indigenous peoples to press for reforms of their own governments often may serve US security better than external US military pressure. Surely military threats against communist governments in Eastern Europe could not have produced such unexpectedly positive results at such low cost as did primarily non-violent demonstrations in 1989. Therefore, to encourage steps in the direction of expanded political accountability, countries such as the United States could more pointedly attempt, through non-violent means such as conditional aid programs and promotion of international human rights covenants, to reform authoritarian regimes and to facilitate the growth of representative government. Although traditional ideas about sovereignty have restricted identities to territorial space, as R.B.J. Walker has pointed out,[68] this confinement need not continue in the future. New security policies can support the growing feeling of individuals that they have responsibility for shaping their governments, a sense that now extends to every continent. The growth of people power, if it continues and remains essentially non-violent, augurs well for the advancement of the other four principles proposed here.

CONCLUSION

In one sense we can look forward to the approaching end of "alternative" security. As more people become aware that there is no genuinely secure alternative to alternative security, then the policies proposed in this volume will surface in mainstream political thinking. As that happens, they will no longer be called *alternative* security policies.

This process is already well underway in some parts of Europe and among intellectual, religious, scientific, and artistic leaders in all countries. Alternative security ideas are spreading not because this is an era of optimism but because the old national security approach is, upon fair-minded reflection, thoroughly discredited. Military instruments and the militarily competitive balance-of-power system are unable to address many, if not all, of the most pressing, long-range security problems of our era. Economic, environmental, human rights, and military questions can no longer be contained in national political arenas. Future success in constructing effective security policies rests on understanding that fragmented identities and stunted psychic development pose the underlying problems that must be overcome to generate the attitudinal, value, and institutional changes that can enable our species to survive, and to survive with dignity.

Species identity and respect for all of creation are the keys to US and world security. With those keys in hand, the present generation is called to close the time-honored door of *national* foreign policy and to open the timeless door of *world* policy. We have reached the historic watershed when it is less important for one's own nation to possess the ability and the right to make war than it is to obtain influence over that ability and that right in the hands of others. To give up the former in order to obtain the latter would be a life-enhancing bargain. That bargain trades the war-making function, now exercised by separate States operating under traditional national sovereignty, for a war-controlling function, to be exercised by many States operating with at least one sovereign function held in common.

Of course, there is nothing inevitable about progress toward the goals of alternative security. Indeed, one clings more to hope for rapid social learning than to confidence based on demonstrated ability to change resistant minds and institutions or to mobilize the hopeless, the indifferent, and the self-centered. Yet, by intentionally seeking to rise above the boundaries that separate nations and maintain the rich-poor divide, by cultivating respect for future generations and for all of nature, and by nurturing the growth and power of psychologically healthy and spiritually sensitive people to shape their own and each other's destinies, surely the development of a principled, peaceful, and compassionate world policy will not exceed our grasp.

NOTES

I would like to thank the Institute for International Peace and Security, University of Notre Dame, and the John D. and Catherine T. MacArthur Foundation for their support in the research and writing of this chapter.

1. George Washington, "Annual State of the Union Address to Congress (January 8, 1790)," in Saxe Commins (ed.), *Basic Writings of George Washington* (New York: Random House, 1948): 567.

2. *See* the chapter by Ralph K. White in this volume: 176, 190-93.

3. The expression is borrowed from William Sloane Coffin's chapter in this volume: 206.

4. Erik H. Erickson, *Gandhi's Truth: On the Origins of Militant Nonviolence* (New York: Norton, 1969): 431-34.

5. *See generally* the chapter by Lloyd J. Dumas in this volume.

6. *Ibid.*: 139-144.

7. *Ibid.*: 143-44, 149-151.

8. Some of these matters are discussed or noted by William Sloane Coffin, Burns H. Weston, and Ralph K. White in their chapters in this volume.

9. People for whom these boundaries are *not* the mainspring of their actions also more frequently find the boundary between military and nonmilitary killing—a line that legitimizes one form of violence and delegitimizes the other—uncompelling.

10. *See generally* Roger Fisher and William Ury, *Getting to Yes: Negotiating Agreement Without Giving In* (Boston: Houghton Mifflin, 1981).

11. This is part of Kenneth Boulding's "chalk theory" of stable peace as noted by Dumas, ch. 5 in this volume: 145.

12. *See* in this connection the words of A. Georgi Arbatov quoted at the outset of the chapter by Bruce Russett in this volume.

13. As the Palme Commission on Disarmament and Security Issues concluded: "Even on the so-called 'conventional' level . . ., war is losing its meaning as an instrument of national policy, becoming instead an engine of senseless destruction that leaves the root causes of conflict unresolved." Palme Commission on Disarmament and Security Issues, *A World at Peace: Common Security in the Twenty-first Century* (Stockholm: The Palme Commission, 1989): 6.

14. For comments on this achievement, see the chapter by Dumas in this volume, ch. 5: 141.

15. *See, e.g.,* George F. Kennan, *The Nuclear Delusion: Soviet-American Relations in the Atomic Age* (New York: Pantheon, 1982): xxviii.

16. Many observers, especially those close to the Reagan and Bush administrations, make the claim that arms control succeeded in eliminating intermediate-range nuclear forces in the INF treaty because of the Reagan Administration's military buildup and tough stand against the Soviet Union. John Lewis Gaddis makes this argument in "Hanging Tough Paid Off," *Bulletin of the Atomic Scientists* 45 (January/February 1989): 11-14. For discussion of this view, see Robert C. Johansen, "Do Preparations for War Increase or Decrease International Security?" in Charles W. Kegley (ed.), *The Long Postwar Peace* (Glenview, IL: Scott, Foresman, forthcoming).

17. Fred C. Iklé and Albert Wohlstetter (Co-chairs), *Report of the Commission on Integrated Long-Term Strategy, Discriminate Deterrence* (Washington: Government Printing Office, 1988). *See also* Zbigniew Brzezinski, "America's New Geostrategy," *Foreign Affairs* 66 (Spring 1988): 680-99.

18. *See* Leonard S. Spector, "We've Forgotten About the Bomb," *New York Times* (November 4, 1989): 25, col. 2.

19. For details, see Stockholm International Peace Research Institute, *SIPRI Yearbook 1988: World Armaments and Disarmament* (New York: Oxford University Press, 1988): 52-58, 101-121.

20. The spread of such technologies to regionally hegemonic countries

exacerbates the problem of uneven growth of power among States, which many theorists long ago concluded is the most fundamental problem of international relations in the contemporary world. The fear that Israel could target Baghdad with missiles, for example, may have contributed to Iraq's decision to modify its Scud-B missiles to extend their range to over 500 miles.

21. For further discussion of this point, see Johansen, "Preparations for War."

22. Several of the analyses in this book explicitly acknowledge the need to focus on the entire war problem, not only nuclear deterrence. *See*, for example, the chapters in this volume by Coffin, ch. 7: 211-13; Warren F. Davis, ch. 2: 44, 46; and Burns H. Weston, ch. 3: 79, 88.

23. For example, Ralph K. White says that minimum deterrence "would seem wise," perhaps even "essential," although mainly for "an intermediate period," ch. 6 in this volume: 184.

24. *See* John Vasquez, "The Deterrence Myth: New Weapons and the Prevention of Nuclear War," in Charles W. Kegley (ed.), *The Long Postwar Peace*.

25. *See, e.g.,* John Mueller, "The Essential Irrelevance of Nuclear Weapons," *International Security*, 13 (Fall 1988): 55-79; and John Mueller, *Retreat from Doomsday: The Obsolescence of Major War* (New York: Basic Books, 1989).

26. *See* White, ch. 6 in this volume: 183-85.

27. It is hardly persuasive to end the time period for analysis at an arbitrary year, 1938, and then claim that deterrence worked when it clearly failed only one year later. *See* White, ch. 6 in this volume: 184.

28. White's additional claim (ch. 6 in this volume: 184) that nuclear weapons kept US-Soviet peace for the past four decades remains unproven, as the Vasquez analysis demonstrates, in the absence of convincing evidence that the two countries would have attacked each other in the absence of nuclear weapons.

29. *Ibid.*

30. Frank Barnaby, former director of the Stockholm International Peace Research Institute, and Egbert Boeker, Professor of Theoretical Physics, Free University of Amsterdam, conclude: "A conventional defensive deterrent is now possible. European security is possible without nuclear weapons. It is impossible with them." *See* Frank Barnaby and Egbert Boeker, "Defence Without Offence: Non-nuclear Defense for Europe," *Peace Studies Paper No. 8* (London: School of Peace Studies, University of Bradford, November 1982).

31. For a provocative discussion of these issues, see Robert Jay Lifton, *The Genocidal Mentality: Nazi Holocaust and Nuclear Threat* (New York, Basic Books, 1990).

32. *See* ch. 3 in this volume: especially 84-96.

33. For discussion of the utility of a permanent UN police force and other institutional innovations for the UN system, see Robert C. Johansen, "The Reagan Administration and the U.N.: The Costs of Unilateralism," *World Policy Journal* 3 (Fall 1986): 601-41.

34. Daniel Yankelovich and Richard Smoke, "America's 'New Thinking,'" *Foreign Affairs* 76 (Fall 1988): 6.

35. *Ibid.*: 9.

36. *See* Weston, ch. 3 in this volume: especially 93-96.

37. *See, e.g.,* Georgi Shakhnazarov, "Governability of the World," *International Affairs*, No. 3 (Moscow, 1988): 16-18, 22. *See also* Shakhnazarov's article in *Pravda* (January 15, 1988): 3.

38. *See* Robert C. Johansen, "United States-Soviet Security Enhancement," *UNESCO Yearbook on Peace and Conflict Studies*, vol. 9 (Westport, CT: Greenwood Press, 1990).

39. *See* the chapter by Bruce Russett, ch. 4 in this volume: 120-24.

40. *See generally* the chapter by Thomas F. Lynch, III, in this volume, ch. 1: especially 9-10.

41. *Ibid.*: 32.

42. *Ibid.*: 35.

43. *See* Paul Kennedy, *The Rise and Fall of the Great Powers: Economic Change and Military Conflict from 1500 to 2000* (New York: Vintage, 1987).

44. John Tirman, *Sovereign Acts: American Unilateralism and Global Security* (New York: Ballinger, 1989): 178.

45. *Ibid.*: 178.

46. Lynch, ch. 1 in this volume: 27-30.

47. *See, e.g.,* Lloyd J. Dumas, *The Political Economy of Arms Reduction: Reversing Economic Decay* (AAAS Selected Symposium No. 80) (Boulder, CO: Westview, 1982); *ibid., The Overburdened Economy* (Berkeley, CA: University of California Press, 1986).

48. The tiny island of Fiji has carried this idea a step further. International peacekeeping has become "the chief task of Fiji's military forces." Ramesh Thakur, *International Peacekeeping in Lebanon: United Nations Authority and Multinational Force* (Boulder, CO: Westview, 1987): 73. Their military personnel consider it a high honor to serve in *ad hoc* UN peacekeeping efforts, where they have performed with courage and distinction.

49. White, ch. 6 in this volume: 199-200.

50. This viewpoint is shared in the chapters in this volume by Weston, ch. 3: 78, 96-97; and Dumas, ch. 5: 153-55, 159-60.

51. World Commission on Environment and Development, *Our Common Future* (New York: Oxford University Press, 1987).

52. Michael G. Renner, "What's Sacrificed When We Arm?" *World Watch* 2 (September-October, 1989): 10.

53. *Ibid.*

54. *See* Hilary F. French "An Environmental Security Council?" *World Watch* 2 (September-October 1989): 7.

55. White, ch. 6 in this volume: 190-93.

56. *See, e.g.,* Gernot Köhler, "Global Apartheid," *Alternatives: A Journal of World Policy* 4 (1978-79).

57. Independent Commission on International Development Issues

(Brandt Commission), *North-South: A Programme for Survival* (Cambridge, MA: MIT Press, 1980): 274.

58. Russett, ch. 4 in this volume, and other studies of democracy support this inference.

59. On accountability in world politics, see Seyom Brown, "Inherited Geopolitics and Emergent Global Realities," in Edward K. Hamilton (ed.), *America's Global Interests* (New York: Norton, 1989): 195-97.

60. George Bush, "When I Talk About a Kinder and Gentler Nation, I Mean It," *New York Times* (October 24, 1988): 5, col. 1.

61. Jessica Tuchman Matthews, "Redefining Security," *Foreign Affairs*, 6, 2 (Spring 1989): 163.

62. For details, see Richard P. Claude and Burns H. Weston (eds.), *Human Rights in the World Community: Issues and Action* (Philadelphia: University of Pennsylvania Press, 1989): ch. 4.

63. Russett, ch. 4 in this volume.

64. Lynch, ch. 1 in this volume: 30-31.

65. *See* Serge Schmemann, "East Germany Let Largest Protests Proceed in Peace," *New York Times* (October 10, 1989): 1, col. 6; Craig R. Whitney, "People's Revolt: Will East Berlin Leaders Survive?" *New York Times* (November 20, 1980): 1, col. 3; Serge Schmemann, "East Germans' New Leader Vows Far-Reaching Reform and Urges an End to Flight," *New York Times* (November 4, 1989): p. 1, col. 6.

66. *See* Celestine Bohlen and Clyde Haberman, "How the Ceausescus Fell: Harnessing Popular Rage," *New York Times* (January 7, 1990): 1, col. 4.

67. *See* Lynch, ch. 1 in this volume: 30-31.

68. R.B.J. Walker, "World Security and International Relations Theory," in Michael Klare and Daniel Thomas (eds.), *Security, Sovereignty, and the Challenge of World Politics* (unpublished manuscript, September 1989): 28-36.

Selected Bibliography

BOOKS AND MONOGRAPHS

Allison, Graham T., Albert Carnesale, and Nye, Joseph S., Jr. (eds.). *Hawks, Doves and Owls: An Agenda for Avoiding Nuclear War.* New York: Norton, 1985.

Alternative Defence Commission. *Defence Without the Bomb.* London: Taylor and Francis, 1984.

_____. *The Politics of Alternative Defence: A Role for a Non-Nuclear Britain.* London: Paladin, 1987.

American Friends Service Committee. *Instead of War: An Inquiry into Nonviolent National Defense.* New York: Grossman, 1967.

Barnaby, Frank, and Egbert Boeker. *Defence Without Offence: Non-Nuclear Defence for Europe.* London: Housmans, 1983.

Barnet, Richard J. *Real Security: Restoring American Power in a Dangerous Decade.* New York: Touchstone/Simon and Schuster, 1981.

Barnet, Richard J., and Richard A. Falk (eds.). *Security in Disarmament.* Princeton, NJ: Princeton University Press, 1965.

Boserup, Anders, and Andrew Mack. *War Without Weapons.* New York: Schocken, 1975.

Boulding, Kenneth E. *Stable Peace.* Austin, TX: University of Texas Press, 1978.

Bowman, Robert M. *Star Wars: Defense or Death Star?* Chesapeake Beach, MD: Institute for Space and Security Studies, 1985.

Boyle, Francis A. *World Politics and International Law.* Durham, NC: Duke University Press, 1985.

_____. *Defending Civil Resistance Under International Law.* Dobbs Ferry, NY: Transnational Publishers, 1987.

Brembeck, Howard S. *The Civilized Defense Plan: Security of Nations Through the Power of Trade.* Fairfax, VA: Hero Books, 1989.

Buzan, Barry. *People, States and Fear: The National Security Problem in International Relations.* Brighton, England: Wheatsheaf Books, 1983.

_____ (ed.). *The International Politics of Deterrence.* London: F. Pinter, 1987.

Carlson, Don, and Craig Comstock (eds.). *Securing Our Planet: How to Succeed When Threats Are Too Risky and There's Really No Defense.* Los Angeles: Tarcher/St. Martin's, 1986.

Carver, Field Lord Marshall. *A Policy for Peace.* London: Faber and Faber, 1982.

Clark, Grenville, and Louis B. Sohn. *World Peace Through World Law: Two Alternative Plans.* Cambridge, MA: Harvard University Press, Third Edition enlarged, 1966.

Claude, Inis, Jr. *Swords Into Plowshares: The Problems and Progress of International Organizations.* New York: Random House, Fourth Edition, 1971.

Cohen, Maxwell, and Margaret E. Gouin. *Lawyers and the Nuclear Debate: Proceedings of the Canadian Conference on Nuclear Weapons and the Law.* Ottawa, Canada: University of Ottawa Press, 1988.

Dean, Jonathan. *Watershed in Europe: Dismantling the East-West Military Confrontation.* Lexington, MA: Lexington Books, 1987.

Derek, Paul. *Defending Europe: Options for Security.* London: Taylor and Francis, 1985.

Deudney, Daniel. *Whole Earth Security: Towards a Geopolitics of Peace.* Washington, DC: Worldwatch Institute, Worldwatch Paper No. 55, July 1983.

Drell, Sidney D. *Facing the Threat of Nuclear Weapons.* Seattle, WA: University of Washington Press, 1983.

Dumas, Lloyd J. *The Political Economy of Arms Reduction: Reversing Economic Decay* (AAAS Selected Symposium No. 80). Boulder, CO: Westview, 1982.

_____. *The Overburdened Economy.* Berkeley, CA: University of California Press, 1986.

Dyson, Freeman. *Weapons and Hope.* New York: Harper and Row, 1984.

Falk, Richard A. *A Study of Future Worlds.* New York: The Free Press, 1975.

_____. *Reviving the World Court.* Charlottesville, VA: University Press of Virginia, 1986.

Fischer, Dietrich. *Preventing War in the Nuclear Age.* Totowa, NJ: Rowman and Allanheld, 1989.

Fischer, Dietrich, Wilhelm Nolte, and Jan Oberg. *Winning Peace: Strategies and Ethics for a Nuclear-Free World.* New York: Crane, Russak, 1989.

Fogg, Richard. *Nonmilitary Defense Against Nuclear Threateners and Attackers.* Stevenson, MD: Center for the Study of Conflict, 1983.

Ford, Daniel F., Henry Kendall, and Steven Nadis. *Beyond the Freeze: The Road to Nuclear Sanity.* Boston: Beacon Press, 1982.

Forsberg, Randall S. *The Case for a Third World Nonintervention Regime.* Brookline, MA: Institute for Defense and Disarmament Studies, Alternative Defense Working Paper No. 6, December 1987.

Forsberg, Randall S., and Rob Leavitt. *Alternative Defense: A New Approach to Building a Stable Peace.* Brookline, MA: Institute for Defense and Disarmament Studies, 1988.

Franck, Thomas M. *Nation Against Nation: What Happened to the U.N. Dream and What the U.S. Can Do About It.* New York: Oxford University Press, 1985.

Freedman, Lawrence. *The Evolution of Nuclear Strategy.* New York: St. Martin's Press, 1981.

Galtung, Johan. *The True Worlds.* New York: The Free Press, 1980.

_____. *There Are Alternatives: Four Roads to Peace and Security.* London: Spokesman, 1984.

Garcia Robles, Alfonso. *Nuclear Disarmament: A Crucial Issue for the Survival of Mankind.* New Delhi: Indian Council for Cultural Relations, 1984.

Generals for Peace and Disarmament. *Reflections on Security in the Nuclear Age: A Dialogue Between Generals East and West.* Moscow: Progress Publishers, 1988.

Geyer, Alan F. *The Idea of Disarmament: Rethinking the Unthinkable.* Washington, DC: Churches' Center for Theology and Public Policy, 1985.

Golden, James R., Asa Clark, and Bruce E. Arlinghaus (eds.). *Conventional Deterrence: Alternatives for European Defense.* Lexington, MA: Lexington Books, 1984.

Halperin, Morton. *Nuclear Fallacy.* Cambridge, MA: Ballinger, 1987.

Hanreider, Wolfram F. *Technology, Strategy, and Arms Control.* Boulder, CO: Westview, 1986.

Harvard Nuclear Study Group. *Living with Nuclear Weapons.* New York: Bantam Books, 1983.

Heilbroner, Robert. *An Inquiry into the Human Prospect.* New York: W. W. Norton, 1974.

Hollins, Harry B., Averill L. Powers, and Mark Sommer. *The Conquest of War: Alternative Strategies for Global Security.* Boulder, CO: Westview, 1989.

Independent Commission on Disarmament and Security Issues (Palme Commission). *Common Security: A Blueprint for Survival.* New York: Simon and Schuster, 1982.

Independent Commission on International Development Issues (Brandt Commission). *North-South: A Programme for Survival.* Cambridge, MA: MIT Press, 1980.

Jasani, Bhupendra, and Toshibomi Sakata (eds.). *Satellites for Arms Control and Crisis Monitoring.* New York: Oxford University Press, 1987.

Jervis, Robert. *Perception and Misperception in International Politics.* Princeton: Princeton University Press, 1976.

Johansen, Robert C. *Towards a Dependable Peace: A Proposal for an Appropriate Security System.* New York: Institute for World Order (now World Policy Institute), 1978.

_____. *The National Interest and the Human Interest: An Analysis of United States Foreign Policy.* Princeton, NJ: Princeton University Press, 1980.

_____. *Towards an Alternative Security System.* New York: World Policy Institute, World Policy Paper No. 24, 1983; *reprinted in* Burns H. Weston (ed.), *Toward Nuclear Disarmament and Global Security: A Search for Alternatives.* Boulder, CO: Westview, 1984: 569-603.

_____. *Toward National Security Without Nuclear Deterrence.* Boston: Explanatory Project on the Conditions of Peace, ExPro Paper No. 8, 1987.

Kaldor, Mary, and Dan Smith (eds.). *Disarming Europe.* London: Merlin, 1982.

Kennan, George F. *The Nuclear Delusion: Soviet-American Relations in the Atomic Age.* New York: Pantheon, 1982.

Keohane, Robert O. *After Hegemony: Cooperation and Discord in the World Political Economy.* Princeton, NJ: Princeton University Press, 1984.

Keohane, Robert O., and Joseph S. Nye, Jr. *Power and Independence: World Politics in Transition.* Boston: Little, Brown, 1977.

Kidd, Jack. *The Strategic Cooperation Initiative.* Charlottesville, VA: Three Presidents Publishing, 1988.

Kothari, Rajni. *Footsteps into the Future: Diagnosis of the Present World and a Design for an Alternative.* New York: The Free Press, 1974.

Kothari, Rajni, Richard Falk, Mary Kaldor, Lim Teck Ghee, Giri Deshingkar, Jimoh Omo-Fadaka, Tamas Szentes, José A. Silva Michelena, Ismael Sabri-Abdalla, and Yoshikazu Sakamoto. *Towards A Liberating Peace.* New York: New Horizons Press, 1988.

Krass, Allan. *Verification: How Much Is Enough?* Lexington, MA: Lexington Books, 1985.

Lall, Betty G. *Prosperity Without Guns: The Economic Impact of Reductions in Defense Spending.* New York: Institute for World Order (now World Policy Institute), 1977.

_____. *Security Without Star Wars.* New York: Council on Economic Priorities, 1987.

Lifton, Robert Jay, and Richard Falk. *Indefensible Weapons: The Psychological and Political Case Against Nuclearism.* New York: Basic Books, 1982.

McNamara, Robert S. *Blundering into Disaster.* New York: Pantheon, 1986.

Mandelbaum, Michael. *The Fate of Nations: The Search for National Security in the 19th and 20th Centuries.* New York: Cambridge University Press, 1988.

_____. *Reconstructing the European Security Order.* New York: Council on Foreign Relations, Critical Issues 1990-1, 1990.

Mazrui, Ali A. *A World Federation of Cultures: An African Perspective.* New York: The Free Press, 1976.

Melman, Seymour. *The Demilitarized Society: Disarmament and Conversion.* Montreal: Harvest House, 1988.

Mendlovitz, Saul H. *On the Creation of a Just World Order.* New York: The Free Press, 1975.

Miller, Arthur Selwyn, and Martin Feinrider (eds.). *Nuclear Weapons and Law.* Westport, CT: Greenwood Press, 1984).

Miller, Steven E., and Stephen Van Evera (eds.). *The Star Wars Controversy.* Princeton, NJ: Princeton University Press, 1986.

Mische, Gerald, and Patrica Mische. *Toward a Human World Order: Beyond the National Security Straitjacket.* New York: Paulist Press, 1977.

Mische, Patricia. *Star Wars and the State of Our Souls.* Minneapolis: Winston Press, 1985.

Mueller, John. *Retreat from Doomsday: The Obsolescence of Major War.* New York: W. H. Freeman, Third Edition, 1989.

National Academy of Sciences. *Nuclear Arms Control: Background and Issues.* Washington, DC: National Academy Press, 1985.

Nye, Joseph S., Jr. *Nuclear Ethics.* New York: The Free Press, 1986.

Office of Technology Assessment. *SDI: Technology, Survivability, and Software.* Washington, DC: U.S. Government Printing Office, 1988.

Osgood, Charles E. *An Alternative to War or Surrender.* Urbana, University of Illinois Press, 1962.

Raskin, Marcus. *Program Treaty for General Security and Disarmament.* Washington, DC: Institute for Policy Studies, 1984.

Rikhye, Indar Jit, Michael Harbottle, and Bjorn Egge. *The Theory and Practice of Peacekeeping.* London: C. Hurst and Co., 1984.

Roberts, Adam. *Civilian Resistance as a National Defense.* Harrisburg, PA: Stackpole, 1968.

_____. *Nations in Arms: Theory and Practice of Territorial Defense.* London: Chatto and Windus, 1976.

Rosecrance, Richard. *The Rise of the Trading State.* New York: Basic Books, 1986.

Russett, Bruce. *The Prisoners of Insecurity: Nuclear Deterrence, the Arms Race, and Arms Control.* San Francisco: W. H. Freeman and Co., 1983.

Schell, Jonathan. *The Abolition.* New York: Knopf, 1984.

Schmid, Alex P. *Social Defence and Soviet Military Power: An Inquiry into the Relevance of an Alternative Defence Concept.* Leiden, The Netherlands: Center for the Study of Conflict, 1985.

Sharp, Gene. *The Politics of Nonviolent Action.* 3 vols. Boston: Porter Sargent, 1974/1980.

_____. *Making the Abolition of War a Realistic Goal.* New York: Institute for World Order (now World Policy Institute), 1981.

_____. *Making Europe Unconquerable: The Potential of Civilian-Based Deterrence and Defense.* Cambridge, MA: Ballinger, 1985.

Smith, Theresa C., and Indu B. Singh. *Security vs. Survival: The Nuclear Arms Race.* Boulder, CO: Lynne Rienner, 1985.

Smoke, Richard. *National Security and the Nuclear Dilemma.* New York: Random House, Second Edition, 1987.

Smoke, Richard, and Willis Harman. *Paths to Peace.* Boulder, CO: Westview, 1987.

Sommer, Mark. *Beyond the Bomb: Living Without Nuclear Weapons — A Field Guide to Alternative Strategies for Building a Stable Peace.* Boston: ExPro Press, 1986.

Stephenson, Carolyn M. (ed.). *Alternative Methods for International Security.* Washington, DC: University Press of America, 1982.

Stockholm International Peace Research Institute. *Policies for Common Security.* Stockholm: SIPRI, 1985.

Thompson, E.P. *Beyond the Cold War.* London: Merlin, 1982.

Thompson, E.P., and Dan Smith (eds.). *Protest and Survive.* New York: Monthly Review Press, 1981.

Tinbergen, Jan. *Revitalizing the United Nations System.* Santa Barbara, CA: Nuclear Age Peace Foundation, Waging Peace Series Booklet No. 13, 1987.

Tinbergen, Jan, and Dietrich Fischer. *Warfare and Welfare: Integrating Security Policy Into Socio-Economic Policy.* New York: St. Martin's Press, 1987.

Tirman, John (ed.). *The Fallacy of Star Wars.* New York: Vintage, 1984.

Tromp, Hylke (ed.). *Non-Nuclear War in Europe: Alternatives for Nuclear Defense.* Groningen, Netherlands: Groningen University Press, 1984.

Tsipis, Kosta, David Hafemeister, and Penny Janeway. *Arms Control Verification: The Technologies that Make it Possible.* Elmsford, NY: Pergamon-Brassey, 1985.

Westing, Arthur H. (ed.). *Global Resources and International Conflict: Environmental Factors in Strategic Policy and Action.* Oxford: Oxford University Press, 1986.

Weston, Burns H. (ed.). *Toward Nuclear Disarmament and Global Security: A Search for Alternatives.* Boulder, CO: Westview, 1984.

Weston, Burns H., Richard A. Falk, and Anthony D'Amato (eds.). *International Law and World Order: A Problem-Oriented Coursebook.* St. Paul, MN: West Publishing, Second Edition, 1990.

White, Ralph K. *Fearful Warriors: A Psychological Profile of U.S.-Soviet Relations.* New York: The Free Press, 1984.

Wilson, Andrew. *The Disarmer's Handbook of Military Technology and Organization.* Hammondsworth, England: Penguin Books, 1983.

Wiseman, Henry (ed.). *Peacekeeping: Appraisals and Proposals.* New York: Pergamon, 1983.

Woolsey, R. James. *Nuclear Arms: Ethics, Strategy, Politics.* San Francisco, I.C.S. Press, 1984.

SYMPOSIA

"Alternative Approaches to Security Policies," *Bulletin of Peace Proposals* 17, 2 (1986).

"Alternative Defense and Security," *Bulletin of Peace Proposals* 4 (1978).

"Common Threats/Common Security," *Peace Review* 1, 2 (Spring 1989).

"European Security in Regional and Global Perspective," *Bulletin of Peace Proposals* 16, 4 (1985).

"The Image of the Enemy: U.S. Views of the Soviet Union," *Journal of Social Issues* 45, 2 (1989).

"A New European Defense," *Bulletin of the Atomic Scientists* 44, 7 (September 1988).

"Non-Provocative Alternative Defense," *Journal of Peace Research* 24, 1 (March 1987).

ARTICLES

Afheldt, Horst. "New Policies, Old Fears," *Bulletin of the Atomic Scientists* 44, 7 (September 1988): 24-28.

Agrell, Wilhelm. "Offensive versus Defensive: Military Strategy and Alternative Defence," *Journal of Peace Research* 24, 1 (1987): 75-85.

Alger, Chadwick F. "Reconstructing Human Polities: Collective Security in the Nuclear Age," in Burns H. Weston (ed.), *Toward Nuclear Disarmament and Global Security: A Search for Alternatives* (Boulder, CO: Westview, 1984): 666-87.

Arbess, Daniel. "The International Law of Armed Conflict in Light of Contemporary Deterrence Strategies: Empty Promises or Meaningful Restraint?" *McGill Law Journal* 30, 1 (December 1984): 89-142.

Boserup, Anders. "A Way to Undermine Hostility," *Bulletin of the Atomic Scientists* 44, 7 (September 1988): 16-19.

Boyle, Francis A. "The Relevance of International Law to the 'Paradox' of Nuclear Deterrence," *Northwestern University Law Review* 80, 6 (Summer 1986): 1407-48.

Brucan, Silviu. "The Establishment of a World Authority: Working Hypotheses," *Alternatives — A Journal of World Policy* 8, 2 (Fall 1982): 209-223; reprinted in Burns H. Weston (ed.), *Toward Nuclear Disarmament and Global Security: A Search for Alternatives*. Boulder, CO: Westview, 1984: 615-28.

Chalmers, Malcolm. "The Case for a European Security Organization," *World Policy Journal* 7, 2 (Spring 1990): 215-50.

Falk, Richard A. "Toward a Legal Regime for Nuclear Weapons," *McGill Law Journal* 28, 3 (July 1983): 519-41.

Ferraro, Vincent, and Kathleen FitzGerald. "The End of a Strategic Era: A Proposal for Minimal Deterrence," *World Policy Journal* 2 (Winter 1984): 339-60.

Fischer, Dietrich, and Alan Bloomgarden. "Non-Offensive Defense," *Peace Review* 1, 2 (Spring 1989): 7-11.

Flanagan, Stephen J. "Nonoffensive Defense Is Overrated," *Bulletin of the Atomic Scientists* 44, 7 (September 1988): 46-48.

Graham, Kennedy. "New Zealand's Non-Nuclear Policy: Towards a Global Security," *Alternatives: Social Transformation and Humane Governance* 12, 2 (April 1987): 217-42.

Johansen, Robert C. "Global Security Without Nuclear Deterrence," *Alternatives: Social Transformation and Humane Governance* 12, 4 (October 1987): 435-60.

――――. "The Reagan Administration and the U.N.: The Costs of Unilateralism," *World Policy Journal* 3 (Fall 1986): 601-41.

Johansen, Robert C., and Saul H. Mendlovitz. "The Role of Law in the Establishment of a New International Order: A Proposal for a

Transnational Police Force," *Alternatives: A Journal of World Policy* 6, 2 (July 1980): 307-37.

Kaldor, Mary. "Disarmament: The Armament Process in Reverse," in E.P. Thompson and Dan Smith (eds.), *Protest and Survive* (New York: Monthly Review Press, 1981): 173-88; *reprinted in* Burns H. Weston (ed)., *Toward Nuclear Disarmament and Global Security: A Search for Alternatives*. Boulder, CO: Westview, 1984: 654-66.

Kothari, Rajni. "Survival in the Age of Transformation," *Gandhi Marg* 4, 2-3 (May-June 1982): 83-106; *reprinted in* Burns H. Weston (ed)., *Toward Nuclear Disarmament and Global Security: A Search for Alternatives*. Boulder, CO: Westview, 1984: 633-54.

Lebow, Richard Ned, and Janice Gross Stein. "Deterrence: The Elusive Dependent Variable," *World Politics* 42, 3 (April 1990): 336-69.

Lutz, Dieter S. "Towards a European Peace Order and a System of Collective Security," *Bulletin of Peace Proposals* 21, 1 (March 1990): 71-76.

Matthews, Jessica Tuchman. "Redefining Security," *Foreign Affairs* 68, 2 (Spring 1989): 162-77.

National Conference of Catholic Bishops. "The Challenge of Peace: God's Promise and Our Response" (Pastoral Letter on War, Armaments and Peace), *Origins* 13, 1 (National Catholic Documentary Service), May 19, 1983.

Nye, Joseph S., Jr. "Arms Control After the Cold War," *Foreign Affairs* 68, 5 (Winter 1989-90): 42-64.

Saperstein, Alvin M. "An Enhanced Non-Provocative Defense in Europe," *Journal of Peace Research* 24, 1 (March 1987): 47-60.

Stephenson, Carolyn. "The Need for Alternative Forms of Security: Crises and Opportunities," *Alternatives: Social Transformation and Humane Governance* 13, 1 (January 1988): 55-76.

Subrahmanyam, K. "Alternative Security Doctrines," *Bulletin of Peace Proposals* 21, 1 (march 1990): 77-85.

Weston, Burns H. "Nuclear Weapons Versus International Law: A Contextual Reassessment," *McGill Law Journal* 28, 3 (July 1983): 541-90.

_____. "Nuclear Weapons and International Law: Prolegomenon to General Illegality," *New York Law School Journal of International and Comparative Law* 2, 4 (1983): 227-56.

_____. "Lawyers and the Search for Alternatives to Nuclear Deterrence," *University of Cincinnati Law Review* 54 (1985): 451-66.

_____. "Law and Alternative Security," *Michigan Journal of International Law* 10, 1 (Winter 1989): 317-32.

Westing, Arthur H. "Towards Eliminating War as an Instrument of Foreign Policy," *Bulletin of Peace Proposals* 21, 1 (March 1990): 29-35.

Woodward, Beverly. "Institutionalization of Nonviolence." *Alternatives: A Journal of World Policy* 3, 21 (August 1977): 49-73.

Zuckerman, Lord Solly. "Reagan's Highest Folly," *New York Review of Books*, April 9, 1987: 35-41.

BIBLIOGRAPHIES

Atkins, Stephen E. *Arms Control and Disarmament, Defense and Military, International Security, and Peace: An Annotated Guide to Sources, 1980-1987.* Santa Barbara, CA: ABC-Clio, 1988.

Whisman, Linda A. "Nuclear Arms Control: A Selected Bibliography." *The International Lawyer,* 18 (Fall 1984): 1077-80.

Acronyms

For a glossary of terms, including acronyms, pertinent to the contents of this volume, see Appendix A in Burns H. Weston (ed.), *Toward Nuclear Disarmament and Global Security: A Search for Alternatives* (Boulder, CO: Westview, 1984).

ABM	Anti-ballistic missile
AMLM	American Military Liaison Mission
ASAT	Anti-satellite
BMD	Ballistic missile defense
CACM	Central American Common Market
CBD	Civilian-based defense
CBMs	Confidence-building measures
CDWB	Conventional deterrence without bounds
CFE	Conventional Forces in Europe
CSBM	Conventional security-building measures
CSCE	Conference on Security and Co-operation in Europe
CTD	Comprehensive test ban
ECOWAS	Economic Community of West African States
EEC	European Economic Community ("Common Market")
GAOR	General Assembly Official Records
GIUK	Greenland-Iceland-United Kingdom
GRIT	Graduated and Reciprocated Initiatives in Tension-reduction
HLG	High-level group
IAEA	International Atomic Energy Agency
IALANA	International Association of Lawyers Against Nuclear Arms
INF	Intermediate-range nuclear forces
ISMA	International satellite monitoring agency
JCS	(US) Joint Chiefs of Staff
JMWG	Joint Military Working Group

JUSTARS	Joint Surveillance Target Attack Radar System
LANAC	Lawyers' Alliance for Nuclear Arms Control
LCNP	Lawyers' Committee on Nuclear Policy
LDCs	Less developed countries
L.N.T.S.	League of Nations Treaty Series
MAD	Mutual assured destruction
MDCs	More developed countries
MFR	Mutual force reduction
MIRVs	Multiple independently targeted reentry vehicles
MNCs	Multinational corporations
MNEs	Multinational enterprises
MPS	Mobile protective shelters
NATO	North Atlantic Treaty Organization
NCA	National Command Authorities
NGOs	Non-governmental Organizations
NOD	Non-offensive defense
NPT	Non-Proliferation Treaty
OAS	Organization of American States
OECD	Organization of Economic Co-operation and Development
OPEC	Organization of Petroleum Exporting Countries
OTH	Over-the-horizon (radar)
PGMs	Precision guided munitions
PRC	People's Republic of China
R&D	Research and development
RPVs	Remotely piloted vehicles
SAC	Strategic Air Command
SAINTS	(US) Satellite interceptors
SAS	Study Group on Alternative Security Planning
SDI	Strategic Defense Initiative ("Star Wars")
SMLM	Soviet Military Liaison Mission
SPD	(West German) Social Democratic Party
START	Strategic Arms Reduction Talks
T.I.A.S.	(US) Treaties and Other International Acts Series
U.N.G.A.	United Nations General Assembly
U.N.T.S.	United Nations Treaty Series
U.S.T.	United States Treaties and Other International Agreements
WTO	Warsaw Treaty Organization ("Warsaw Pact")

Index

About the Book

Alternative Security offers the thinking person a place to begin to kick the "nuclear habit." Even as it accepts the premise that war is endemic to the human condition, it provides reassurance that an other-than-nuclear deterrence policy can work to effectively safeguard national and transnational interests.

These eight original essays, accompanied by a contextual introduction and a selected bibliography, constitute an interdisciplinary approach to solving the problems of deterrence. Experts with distinct perspectives—military, technological, legal, economic, political, psychological, and spiritual—have collaborated to present realistic alternatives to policies based on nuclear threat.

Principled and pragmatic, *Alternative Security* is ideal for both students and policymakers. It is especially timely in its response to new possibilities being created by profound changes in Eastern Europe and, indeed, the world over.